Hayk Sedrakyan
Nairi Sedrakyan

AMC 10 preparation book

American Mathematics Competitions
preparation book

2021

About RSM

According to Russian tradition - the study of mathematics is the preeminent tool of mental development, and of learning to think powerfully. The top academic minds of the Soviet Union were tasked with developing a right way - a curriculum and methodology that would lead generations of students to their full potential. The resulting methods and textbooks came to be **used by elite schools** globally including in China, India, and Europe.

In time, the Russian methodology gave rise to generations of thinkers with a deep mathematical foundation, who could think critically, creatively, logically, and who welcomed challenge and the unknown. The result was having one of the strongest math schools in the world, standing next to the worlds best math schools, such as French or German ones.

The Russian School of Mathematics (RSM) was opened by two such immigrant women, who disappointed by the level of math education in the United States, opened a school for their own children and the children of their community. The curriculum and methodology, perfected over 20 years by a team of gifted academics, inspired by elite mathematical schools of the former Soviet Union, **adapted for the American environment**.

Today, RSMs award-winning, after-school math enrichment program serves about 40,000 K-12 students across North America. Ranked among the **top 10** schools with our students some of the *brightest young people in the world* by the the John Hopkins University Center for Talented Youth, RSM helps students develop a **deep and life-long understanding of mathematics**, as well as an advanced way of thinking and approaching problems.

In 2017, in order to develop and create a top-level math competition preparation program and math competition curriculum for RSM, the authors have started a collaboration with the newly opened math competitions department of RSM, Boston, USA. Now, this department includes hundreds of strongest students and dozens of math competition coaches. The authors would like to thank RSM for the support during the publication of this book.

Russian School of Mathematics 2021 ©

About the authors

Hayk Sedrakyan is an IMO medal winner, professional mathematical Olympiad coach in greater Boston area, Massachusetts, USA. He is the Dean of math competition preparation department at RSM. He has been a Professor of mathematics in Paris and has a PhD in mathematics (optimal control and game theory) from the UPMC - Sorbonne University, Paris, France. Hayk is a Doctor of mathematical sciences in USA, France, Armenia and holds three master's degrees in mathematics from institutions in Germany, Austria, Armenia and has spent a small part of his PhD studies in Italy. Hayk Sedrakyan has worked as a scientific researcher for the European Commission (sadco project) and has been one of the Team Leaders at Harvard-MIT Mathematics Tournament (HMMT). He took part in the International Mathematical Olympiads (IMO) in United Kingdom, Japan and Greece. Hayk has been elected as the President of the students general assembly and a member of the management board of the *Cite Internationale Universitaire de Paris* (10,000 students, 162 different nationalities) and the same year they were nominated for the Nobel Peace Prize.

Nairi Sedrakyan is involved in national and international mathematical Olympiads having been the President of Armenian Mathematics Olympiads and a member of the IMO problem selection committee. He is the author of the most difficult problem ever proposed in the history of the International Mathematical Olympiad (IMO), 5th problem of 37th IMO. This problem is considered to be the hardest problems ever in the IMO because none of the members of the strongest teams (national Olympic teams of China, USA, Russia) succeeded to solve it correctly and because national Olympic team of China (the strongest team in the IMO) obtained a cumulative result equal to 0 points and was ranked 6th in the final ranking of the countries instead of the usual 1st or 2nd place. The British 2014 film X+Y, released in the USA as *A Brilliant Young Mind*, inspired by the film *Beautiful Young Minds* (focuses on an English mathematical genius chosen to represent the United Kingdom at the IMO) also states that this problem is the hardest problem ever proposed in the history of the IMO (minutes 9:40-10:30). Nairi Sedrakyan's students (including his son Hayk Sedrakyan) have received 20 medals in the International Mathematical Olympiad (IMO), including Gold and Silver medals.

Overview

This book consists **only of author-created problems with author-prepared solutions** and it is intended as a teacher's manual of mathematics, a self-study handbook for high-school students and mathematical competitors interested in American Mathematics Competitions (especially AMC 10). The book teaches problem solving strategies and aids to improve problem solving skills. The book includes a list of the most useful theorems and formulas for AMC 10, it also includes 12 sets of author-created AMC 10 type practice tests (300 author-created AMC 10 type problems and their detailed solutions). National Math Competition Preparation (NMCP) program of RSM used part of these 12 sets of practice tests to train students for AMC 10, as a result **75 % of NMCP high school students qualified for AIME**. The authors provide both a list of answers for all 12 sets of author-created AMC 10 type practice tests and author-prepared solutions for each problem.

Keywords: AMC 10 preparation book, American Mathematics Competitions preparation, AMC problems and solutions sample tests, AMC 10 problems and solutions sample tests, AMC 10 questions sample tests, AMC problem solving strategies.

In case of any comments please contact sedrakyan.hayk@gmail.com

Mathematical competition is not about winning or losing; it is about mastering the art of thinking creatively and smart.
The true effectiveness of math competition training is to develop creative thinking and problem solving skills that will help in future careers.
Hayk Sedrakyan.

Contents

1 List of the most useful formulas and theorems for AMC 10   **12**
- 1.1 Algebra (AMC 10): the most useful formulas and theorems 12
  - 1.1.1 Basic factorization formulas and binomial expansions. 12
  - 1.1.2 Quadratic and cubic equations, Vieta's formulas. 13
  - 1.1.3 Bézout's little theorem, linear factorization, Newton's divided difference formula. . 14
  - 1.1.4 Formulas for arithmetic, geometric and Fibonacci sequences 15
  - 1.1.5 HM-GM-AM-QM inequalities. 17
  - 1.1.6 Cauchy-Bunyakovsky-Schwarz inequality. 17
  - 1.1.7 Sedrakyan's inequality. 17
  - 1.1.8 Sedrakyan's power sums triangle. 18
  - 1.1.9 Cartesian coordinate system, some important formulas. 19
- 1.2 Geometry (AMC 10): the most useful formulas and theorems 21
  - 1.2.1 Formulas for plane shapes. 21
  - 1.2.2 Tangential, cyclic, bicentric and extangential quadrilaterals, Pitot's theorem. . . 22
  - 1.2.3 Heron's formula and Brahmagupha's formula. 23
  - 1.2.4 Ceva's theorem, Menelaus' theorem, Stewart's theorem, Ptolemy's theorem. . . 24
  - 1.2.5 Bretschneider's formula, *diagonals and sides* area formula of a quadrilateral. . . 24
  - 1.2.6 Parameshvara's formula for circumradius. 25
  - 1.2.7 Formulas for volume and surface areas of three-dimensional shapes. 26
  - 1.2.8 Trigonometric identities. 27
  - 1.2.9 Complex numbers, de Moivre's formula, Euler's formula. 29
- 1.3 Number theory (AMC 10): the most useful formulas and theorems 30
  - 1.3.1 Unique-prime-factorization theorem (fundamental theorem of arithmetic). . . . 30
  - 1.3.2 Number of divisors of a composite number, sum and product of divisors. 30
  - 1.3.3 One useful lemma. 30
  - 1.3.4 Modular arithmetic and congruence relation. 31
  - 1.3.5 Fermat's little theorem and Wilson's theorem. 31
- 1.4 Combinatorics and probability (AMC 10): the most useful formulas and theorems . . . 32
  - 1.4.1 Rule of sum and rule of product. 32
  - 1.4.2 Permutations. 32
  - 1.4.3 Combinations. 32
  - 1.4.4 Stars and bars technique (integer equations). 33
  - 1.4.5 Probability. 33

2 AMC 10 type practice tests   **35**
- 2.1 AMC 10 type practice test 1 35
- 2.2 AMC 10 type practice test 2 39
- 2.3 AMC 10 type practice test 3 42
- 2.4 AMC 10 type practice test 4 45
- 2.5 AMC 10 type practice test 5 48

	2.6	AMC 10 type practice test 6	51
	2.7	AMC 10 type practice test 7	54
	2.8	AMC 10 type practice test 8	58
	2.9	AMC 10 type practice test 9	62
	2.10	AMC 10 type practice test 10	65
	2.11	AMC 10 type practice test 11	68
	2.12	AMC 10 type practice test 12	71

3 Answers — **74**

    3.1 Answers of AMC 10 type practice tests 74

4 Solutions — **77**

    4.1 Solutions of AMC 10 type practice test 1 77
    4.2 Solutions of AMC 10 type practice test 2 90
    4.3 Solutions of AMC 10 type practice test 3 101
    4.4 Solutions of AMC 10 type practice test 4 112
    4.5 Solutions of AMC 10 type practice test 5 121
    4.6 Solutions of AMC 10 type practice test 6 133
    4.7 Solutions of AMC 10 type practice test 7 144
    4.8 Solutions of AMC 10 type practice test 8 159
    4.9 Solutions of AMC 10 type practice test 9 171
    4.10 Solutions of AMC 10 type practice test 10 184
    4.11 Solutions of AMC 10 type practice test 11 197
    4.12 Solutions of AMC 10 type practice test 12 210

Acknowledgment

The authors would like to thank their family for the support.

Introduction

What is AMC 10?

The American Mathematics Competitions (AMC) are math competitions in middle and high school, organized by the Mathematical Association of America (MAA), that begin the multy-level selection process of the United States team for the International Mathematical Olympiad (IMO).
There are three levels: **AMC 8** (for students in grade 8 or below), **AMC 10** (for students in grade 10 or below), **AMC 12** (for students in grade 12 or below).
The track usually starts with AMC 8. It serves as a practice to prepare for AMC 10 and AMC 12.
Students who took part in AMC 10 and were in the top 2.5 percent are invited to take part in the American Invitational Mathematics Examination (AIME).
Students who took part in AMC 12 and were in the top 5 percent are also invited to take part in the American Invitational Mathematics Examination (AIME).
Students who qualify through AMC 10 to take part in AIME and perform well enough on AIME are then invited to the United States of America Junior Mathematical Olympiad (USAJMO).
Students who qualify through AMC 12 to take part in AIME and perform well enough on AIME are then invited to the United States of America Mathematical Olympiad (USAMO).
Qualifying for USAJMO or USAMO is widely considered as one of the most prestigious awards for high school students in the United States.
Top 30 performing students on USAJMO or USAMO are invited to go to the Mathematical Olympiad Summer Program (MOSP or MOP) with the goal of providing them a deep foundation in math Olympiads.
Top 12 performing students are invited to take the Team Selection Test (TST).
Top 6 performing students are selected from these top 12 students to form the United States International Math Olympiad team (US IMO team).

What benefits does AMC participation give to students?

There are many benefits for taking part and doing well on AMC. The list of top scoring students becomes available to colleges, institutions and programs interested in attracting students with strong math background. This gives the applicant an advantage over the other applicants during the admission process. Moreover, sometimes even the best US (or international) colleges take these achievements into consideration and offer a full study scholarship.
Besides this, top scorers on AMC 10 and AMC 12 qualify to participate in the next rounds of math competitions, leading to become a team member of the United States team to take part in the International Mathematical Olympiad (IMO). The most prestigous math Olympiad in the world.
One of the most important benefits is that mathematical competitions are not about winning or losing, they are about mastering the art of thinking creatively and smart. The true effectiveness of math competition training is to develop creative thinking and problem solving skills that will help in future careers.

Strategy advices: becoming more strategic

Strategic decision 1. AMC 10 is a math competition where the problems change anually, but the list of possible topics does not change. Taking this into consideration, the authors provide in the book a list of the most useful formulas and theorems for AMC 10. Mastering this list of formulas and theorems plays a crucial role for student's performance during the actual AMC 10.

Strategic decision 2. The authors aimed to create AMC 10 type practice tests in order to help students to get prepared for AMC 10. **All problems in the book are author-created problems with author-prepared solutions.** They made the tests as close as possible to the topics of AMC 10, but intentionally made most of the tests *slightly* harder than actual AMC 10 tests, believing that if students train to solve slightly harder problems, then it will be easier for them to solve AMC 10 problems during the actual competition. As AMC 10 is 75 minutes math competition with 25 problems of increasing difficulty, this type of preparation also helps to save some time during the actual AMC 10 competition and avoid the possibility of running out of the time.

Strategic decision 3. The best strategy for AMC 10 is to solve all 25 problems correctly within given 75 minutes. This is not always the case and sometimes students need to follow certain strategy to get qualified to AIME (American Invitational Mathematics Examination). With the strategy described in this paragraph, getting qualified to AIME through AMC 12 becomes easier than getting qualified to AIME through AMC 10, even for a ninth grader. AIME cutoff score differs from year to year, as those who rank in the top 5% on AMC 12 get qualified to AIME and those who rank in the top 2.5% on AMC 10 get qualified to AIME. If student's main goal is to get qualified to AIME, then one needs to understand AMC 10 and AMC 12 grading systems. Both are multiple choice math competition tests, where correct answers are worth 6 points, incorrect answers are worth 0 points and unanswered questions are worth 1.5 points. From year 2000 to year 2020 AIME cutoff scores (through AMC 12) were from 90 to 100, and (through AMC 10) were from 100 to 120. This means that if a student manages to solve at least 17 problems correctly in AMC 12, then the student gets qualified to AIME, as $17 \cdot 6 = 102$. Scoring at least 120 in AMC 10, means solving correctly about 20 problems. Solving at least 17 problems correctly may be challenging, often students run out of time and try to guess the answers. Students who realize that they are going to run out of time, their strategy to get qualified to AIME can be using the advantages of the scoring system of AMC 12, for example they can concentrate on the easiest 14 problems of the test, solve them correctly and do not answer the other 11 problems. So, they have 75 minutes to solve the easiest 14 problems and they will get $11 \cdot 1.5 = 16.5$ points for skipping 11 problems. Their total score will be $14 \cdot 6 + 11 \cdot 1.5 = 100.5$ and they can get qualified to AIME. Note that every year about 10 problems are exactly the same in AMC 10 and AMC 12 tests. This, means that if the student uses this strategy getting qualified to AIME through AMC 12 gets easier than through AMC 10, as talking roughtly during AMC 12 student needs to solve 14 problems of the same difficulty as in AMC 10, but AIME cutoff through AMC 12 is lower. We do **not** encourage this strategy, as long as the **only** goal of the student is to get qualified to AIME and not to go further to USAMO.

Strategic decision 4. US(J)AMO qualificaition is determined based on AMC 10 score plus AIME score. In the previous paragraph, we have already explained AMC 10 grading system. AIME is 3 hours math competition with 15 problems of increasing difficulty, such that each answer is an integer number between 0 and 999 (both 0 and 999 inclusive). Each correct answer is scored as one point, each incorrect or blank answer is scored as zero point. Therefore, AIME final score is an integer number from 0 to 15 (both 0 and 15 inclusive). To determine eligibility for the US(J)AMO student's score on AMC 10 (or AMC 12) is added to 10 times the score on AIME. For example, if a student scores 120 on AMC 10 (or AMC 12) and manages to solve 10 problems correctly on AIME, then student's final score is $120 + 10 \cdot 10 = 220$. The cutoff for getting qualified to US(J)AMO is usually from 220 to 230 combined points. Taking this into consideration, if your goal is not only to get qualified to USAMO, but also to go further to IMO, then we encourage you not to use the strategy described in the previous paragraph. In this case, simply try to do your best both in AMC 10 (or AMC 12) and in AIME in order to score as high as possible (the most diserable is at least 230 combined points).

Strategic decision 5. Do not worry and **keep in mind that mathematical competitions are not about winning or losing, they are about mastering the art of thinking creatively and smart**. So, no matter whether the student gets qualified to AIME or not, with proper math competition training (in the end) the student always wins. Good luck.

Chapter 1

List of the most useful formulas and theorems for AMC 10

As a reference the authors would like to provide a list of the most useful formulas and theorems for AMC 10. This list is very useful and important not only for AMC 10, but also for AMC 12, AIME, USAMO, IMO and various mathematical competitions. Taking into consideration that the main topics of AMC 10 are algebra, geometry, number theory, combinatorics and probability, we will divide this chapter into 4 sections, respectively.

1.1 Algebra (AMC 10): the most useful formulas and theorems

1.1.1 Basic factorization formulas and binomial expansions.

$a^2 - b^2 = (a - b)(a + b)$.
$(a - b)^2 = a^2 - 2ab + b^2$.
$a^2 + b^2 = (a - b)^2 + 2ab$.
$(a + b)^2 = a^2 + 2ab + b^2$.
$a^2 + b^2 = (a + b)^2 - 2ab$.
$(a^2 + b^2)(c^2 + d^2) = (ac + bd)^2 + (ad - bc)^2$.
$(a^2 + b^2)(c^2 + d^2) = (ac - bd)^2 + (ad + bc)^2$.
$(a + b + c)^2 = a^2 + b^2 + c^2 + 2(ab + bc + ac)$.

$a^3 - b^3 = (a - b)(a^2 + ab + b^2)$.
$a^3 + b^3 = (a + b)(a^2 - ab + b^2)$.
$(a - b)^3 = a^3 - 3a^2b + 3ab^2 - b^3$.
$a^3 - b^3 = (a - b)^3 + 3ab(a - b)$.
$(a + b)^3 = a^3 + 3a^2b + 3ab^2 + b^3$.
$a^3 + b^3 = (a + b)^3 - 3ab(a + b)$.
$a^3 + b^3 + c^3 - 3abc = (a + b + c)(a^2 + b^2 + c^2 - ab - bc - ac)$.
$a^3 + b^3 + c^3 - 3abc = \frac{1}{2}(a + b + c)\big((a - b)^2 + (b - c)^2 + (c - a)^2\big)$.
$(a + b + c)^3 = a^3 + b^3 + c^3 + 3(a + b)(b + c)(c + a)$.

$ab + a + b + 1 = (a + 1)(b + 1)$, mostly used in algebra and number theory problems where a and b are integers (for example, when we deal with divisibility problems).
$ab - a - b + 1 = (a - 1)(b - 1)$, mostly used in algebra and number theory problems where a and b are integers (for example, when we deal with divisibility problems).

1.1.2 Quadratic and cubic equations, Vieta's formulas.

Given a quadratic equation of a general form $ax^2 + bx + c = 0$, where x is an unknown, a, b, c are given coefficients and $a \neq 0$. Note that, any quadratic equation has at most two solutions. If there is no real solution, there are two complex solutions. If there are two coinciding solutions, then it is called a *double root*. $D = b^2 - 4ac$ is called *discriminant*.

If the discriminant is 0, then quadratic equation has one real solution.
If the discriminant is negative, then quadratic equation has no real solutions.
If the discriminant is positive, then quadratic equation has two real solutions.
These two real solutions x_1, x_2 are given by the following formulas:

$$x_1 = \frac{-b - \sqrt{b^2 - 4ac}}{2a},$$

$$x_2 = \frac{-b + \sqrt{b^2 - 4ac}}{2a}.$$

Factoring quadratic expressions. Note that in this case quadratic equation can be factored in the following way:
$$ax^2 + bx + c = a(x - x_1)(x - x_2).$$

Completing the square method. Note that, using the formula $A^2 + 2AB + B^2 = (A+B)^2$, we obtain

$$ax^2 + bx + c = a\left(x^2 + \frac{b}{a}x + \frac{c}{a}\right) = a\left(x^2 + \frac{b}{a}x + \left(\frac{b}{2a}\right)^2 - \left(\frac{b}{2a}\right)^2 + \frac{c}{a}\right) = a\left(\left(x + \frac{b}{2a}\right)^2 - \frac{b^2 - 4ac}{4a^2}\right).$$

Thus, it follows that
$$ax^2 + bx + c = a\left(x + \frac{b}{2a}\right)^2 - \frac{b^2 - 4ac}{4a}.$$

This method is called *completing the square* and it can be used for different purposes. For example, for finding the minimum and maximum value of a quadratic equation.

Minimum and maximum value of a quadratic expression. Minimum and maximum value of a quadratic expression $f(x) = ax^2 + bx + c, a \neq 0$ can be found in different ways, for example by completing the square, graphically, using derivatives. Let us consider the following two cases:

Case 1. If $a > 0$, then the maximum value is infinite and from completing of the square form we obviously see that the minimum value occurs at $x = -\frac{b}{2a}$, (as any quadratic equation graphically represents a parabola, this point is also called the vertex of corresponding parabola open upward).
Note that in this case the minimum value of the quadratic equation is equal to:

$$f\left(-\frac{b}{2a}\right) = \frac{4ac - b^2}{4a}.$$

Case 2. If $a < 0$, then the minimum value is infinite and from completing of the square form we obviously see that the maximum value occurs at $x = -\frac{b}{2a}$, (at the vertex of the corresponding parabola open downward).
Note that in this case the maximum value of the quadratic equation is equal to:

$$f\left(-\frac{b}{2a}\right) = \frac{4ac - b^2}{4a}.$$

Vieta's formulas are named after a prominent French mathematician François Viète. We provide Vieta's formulas only for quadratic and cubic equations, as they are used widely in different math competitions.

Note that Vieta's formula holds true for a polynomial of any finite power.

Vieta's formula for a quadratic equation of a general form $ax^2 + bx + c = 0$, where $a \neq 0$, states that the sum of the roots (solutions) x_1, x_2 is equal to $-\dfrac{b}{a}$ and the product of the roots is equal to $\dfrac{c}{a}$. This can be rewritten in the following way:

$$ax^2 + bx + c = 0, a \neq 0,$$

$$\begin{cases} x_1 + x_2 = -\dfrac{b}{a}, \\ x_1 \cdot x_2 = \dfrac{c}{a}. \end{cases}$$

In the case of the cubic equation, we have that

$$ax^3 + bx^2 + cx + d = 0, a \neq 0,$$

$$\begin{cases} x_1 + x_2 + x_3 = -\dfrac{b}{a}, \\ x_1 \cdot x_2 + x_2 \cdot x_3 + x_1 \cdot x_3 = \dfrac{c}{a}, \\ x_1 \cdot x_2 \cdot x_3 = -\dfrac{d}{a}. \end{cases}$$

1.1.3 Bézout's little theorem, linear factorization, Newton's divided difference formula.

Bézout's little theorem, also called the *polinomial remainder theorem* is named after a French matematician *Étienne Bézout*. Before giving the formulation of the theorem, let us give the defition of *Euclidean division of polynomials*, also called *long division of polynomials*.

Euclidean division of polynomials (long division of polynomials). Let P and D be two polynomials, where D is not zero, then there exist polynomials Q (called a *quotient*) and R (called a *remainder*), such that

$$P = D \cdot Q + R,$$

where either $R = 0$ or the degree of polynomial R is less than the degree of polynomial D (denoted as $\deg(R) < \deg(D)$). Here P is called *the dividend* and D is called *the divisor*. Moreover, Q and R are uniquely defined.

Bézout's little theorem. The remainder of the division of a polynomial $P(x)$ by a linear polynomial $x - k$ is equal to $P(k)$.

This theorem is named after a French mathematician *Étienne Bézout*.

Factor theorem. A polynomial $P(x)$ has a factor $(x - k)$ if and only if $P(k) = 0$ (k is a root of polynomial $P(x)$).

Linear factorization theorem. Let n be a nonnegative integer and

$$P(x) = a_n x^n + a_{n-1} x^{n-1} + ... + a_1 x + a_0,$$

be a polynomial of degree n in variable x, where $a_n, a_{n-1}, ..., a_1, a_0$ are the coefficients of the polynomial and $a_n \neq 0$ (a_n is called the leading coefficient and a_0 is called the constant term). Then $P(x)$ is possible to represent as the product of n linear factors

$$P(x) = a_n(x - x_1)(x - x_2) \cdot ... \cdot (x - x_n),$$

where $x_1, x_2, ..., x_n$ are the roots of polynomial $P(x)$.

Remark. Note that some of the values of x_i can be complex, where $i = 1, 2, ..., n$.

Lemma (consequence of Newton's divided difference formula). Let $P(x)$ be a polynomial of degree n in variable x. Let $t_1, t_2, ..., t_n$ be any numbers, then there exist unique numbers $c_0, c_1, ..., c_n$, such that

$$P(x) = c_0 + c_1(x - t_1) + c_2(x - t_1)(x - t_2) + ... + c_n(x - t_1)(x - t_2)...(x - t_n).$$

Note that we modified and wrote in a simpler way the classical formulation of *Newton's divided difference formula* to be able to apply it to AMC problems (if needed). It is named after a prominent English mathematician, physicist and astronomer *Isaac Newton*.

Remark. Stating the values of numbers $c_0, c_1, ..., c_n$ is outside of the scope of this book, as their values are not used in AMC problems, only this representation is sometimes used in AMC problems. Nevertheless, the classical formulation of Newton's divided difference formula used in the literature provides formulas for the values of numbers $c_0, c_1, ..., c_n$.

1.1.4 Formulas for arithmetic, geometric and Fibonacci sequences

Definition (arithmetic sequence). A sequence of numbers such that the difference between the consecutive terms is constant number is called an *arithmetic sequence*.

This constant number is called the common difference and usually is denoted by d. For a positive integer n we denote n-th term of this sequence by a_n. Therefore, this definition means that

$$a_{n+1} = a_n + d,$$

where $n = 1, 2, ...$.

The sum of the first n terms is usually denote by S_n, that is $S_n = a_1 + a_2 + ... + a_n$. We provide the following two important formulas related to arithmetic sequences.

$$a_n = a_1 + (n-1)d,$$

and

$$S_n = \frac{a_1 + a_n}{2} \cdot n.$$

Definition (geometric sequence). A sequence of numbers where each term after the first is found by multiplying the previous one by a non-zero constant number is called a *geometric sequence*.

This constant number is called *common ratio* and usually is denoted by r. For a positive integer n we denote n-th term of this sequence by g_n. Therefore, this definition means that

$$g_{n+1} = g_n \cdot r,$$

where $n = 1, 2, ...$.

The sum of the first n terms is usually denote by S_n, that is $S_n = g_1 + g_2 + ... + g_n$. We provide the following two important formulas related to geometric sequences.

$$g_{n+1} = g_1 \cdot r^n,$$

and

$$S_n = \frac{g_1(1 - r^n)}{1 - r},$$

where $r \neq 1$.
If $r = 1$, then

$$S_n = n \cdot g_1.$$

Infinite geometric sequences. The sum of all terms of an infinitie geometric sequence is denoted by
$$S_\infty = g_1 + g_2 + ... + g_n + ...,$$
and we take $-1 < r < 1$. We provide the following important formula for the sum of an infinite geometric sequence, where $-1 < r < 1$.
$$S_\infty = \frac{g_1}{1-r}.$$

Definition (Fibonacci sequence). A sequence of numbers where each number is the sum of the two preceding ones, starting from 0 and 1, is called the *Fibonacci sequence*.

It is named so after an italian mathematician *Leonardo Bonacci* also known as *Fibonacci*. The terms of this sequence are called *Fibonacci numbers* and are denoted by F_n, for a nonnegative integer n.

According to the definition, we have that $F_0 = 0$, $F_1 = 1$ and for $n \geq 2$ it follows that
$$F_n = F_{n-1} + F_{n-2}.$$

Let us provide first few terms of the Fibonacci sequence: 0, 1, 1, 2, 3, 5, 8, 13, 21, 34, 55, 89, ...

Remark. In the literature, sometimes you can see that F_0 is ommited and the sequence starts with $F_1 = F_2 = 1$.

1.1.5 HM-GM-AM-QM inequalities.

The following inequalities are called the HM (harmonic mean)-GM (geometric mean)-AM (arithmetic mean)-QM (quadratic mean) inequalities. At first, let us formulate the HM-GM-AM-QM inequalities for two positive real numbers a and b, that is

$$\frac{2}{\frac{1}{a}+\frac{1}{b}} \leq \sqrt{ab} \leq \frac{a+b}{2} \leq \sqrt{\frac{a^2+b^2}{2}}.$$

Note that the equality holds true if and only if $a = b$.
Now, let us formulate the HM-GM-AM-QM inequalities for any n positive real numbers $a_1, a_2, ..., a_n$, that is

$$\frac{n}{\frac{1}{a_1}+\frac{1}{a_2}+...+\frac{1}{a_n}} \leq \sqrt[n]{a_1 \cdot a_2 \cdot ... \cdot a_n} \leq \frac{a_1+a_2+...+a_n}{n} \leq \sqrt{\frac{a_1^2+a_2^2+...+a_n^2}{n}}.$$

Note that the equality holds true if and only if $a_1 = a_2 = ... = a_n$.

1.1.6 Cauchy-Bunyakovsky-Schwarz inequality.

Cauchy-Bunyakovsky-Schwarz inequality is considered to be one of the most important inequalities in mathematics. It is named after a prominent French mathematician *Augustin-Louis Cauchy*, a Russian mathematician *Viktor Bunyakovsky* and a Prussian mathematician *Karl Hermann Amandus Schwarz*.

Cauchy-Bunyakovsky-Schwarz inequality. For any real numbers $a_1, a_2, ..., a_n, b_1, b_2, ..., b_n$, we have

$$(a_1^2 + ... + a_n^2)(b_1^2 + ... + b_n^2) \geq (a_1 b_1 + ... + a_n b_n)^2.$$

1.1.7 Sedrakyan's inequality.

The following inequality is known as *Sedrakyan's inequality* (published in 1997), it is also called *Engel's form* (published in 1998) or *Titu's lemma* (published in 2003).

Sedrakyan's inequality. For any real numbers $a_1, a_2, ..., a_n$ and positive real numbers $b_1, b_2, ..., b_n$, we have

$$\frac{a_1^2}{b_1} + \frac{a_2^2}{b_2} + ... + \frac{a_n^2}{b_n} \geq \frac{(a_1 + a_2 + ... + a_n)^2}{b_1 + b_2 + ... + b_n}.$$

Moreover, the equality holds true when $\frac{a_1}{b_1} = \frac{a_2}{b_2} = ... = \frac{a_n}{b_n}$.

This inequality helps to prove very hard fractional inequalities in a straightforward and relatively simple way. It can be also used to solve fractional equations. Interested reader can find different generalizations of this inequality in the book *Algebraic Inequalities* of *Hayk Sedrakyan* and *Nairi Sedrakyan* (published by Springer).

Example: Application of Sedrakyan's inequality (IMO 1995, Problem 2). Let a, b, c be positive real numbers such that $abc = 1$. Prove that

$$\frac{1}{a^3(b+c)} + \frac{1}{b^3(c+a)} + \frac{1}{c^3(a+b)} \geq \frac{3}{2}.$$

Proof. Using Sedrakyan's inequality we are able to provide *one-line proof* for this IMO problem. Note that

$$\frac{\left(\frac{1}{a}\right)^2}{a(b+c)} + \frac{\left(\frac{1}{b}\right)^2}{b(c+a)} + \frac{\left(\frac{1}{c}\right)^2}{c(a+b)} \geq \frac{\left(\frac{1}{a}+\frac{1}{b}+\frac{1}{c}\right)^2}{2(ab+bc+ac)} = \frac{ab+bc+ac}{2} \geq \frac{3\sqrt[3]{a^2b^2c^2}}{2} = \frac{3}{2}.$$

1.1.8 Sedrakyan's power sums triangle.

In mathematical competitions we often deal with sums of the forms $1^k + 2^k + ... + n^k$, where n and k are positive integers. For $k = 1$ almost everyone remembers the formula $1 + 2 + ... + n = \dfrac{n(n+1)}{2}$, but very often in order to solve algebra or number theory problems students need to use the corresponding formulas for $1^2 + 2^2 + ... + n^2$, $1^3 + 2^3 + ... + n^3$, $1^4 + 2^4 + ... + n^4$ or higher powers. *Hayk Sedrakyan's* idea was to create a simple and self-constructive *Pascal-type triangle* for sums of powers.

Sedrakyan's power sums triangle. At first, let us provide the triangle and afterward explain how the self-constructive principle of this triangle works. This work was published in the book *Algebraic Inequalities* of *Hayk Sedrakyan* and *Nairi Sedrakyan* (published by Springer).

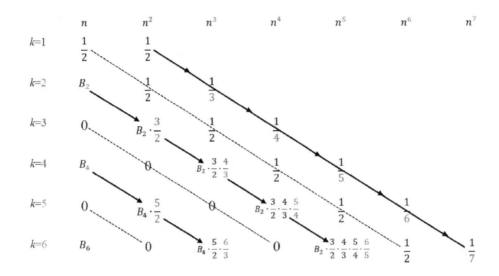

Self-construction principle. Almost all students remember the following well-known formula $1 + 2 + ... + n = \dfrac{n(n+1)}{2}$, we rewrite this formula as $1 + 2 + ... + n = \dfrac{1}{2} \cdot n + \dfrac{1}{2} \cdot n^2$. The first line of the considered *triangle* consists of these two coefficients $\left(\dfrac{1}{2}, \dfrac{1}{2}\right)$ and as in the case of *Pascal's triangle* using the coefficients of the first line we construct the next lines in order to find the value of the sum $1^k + 2^k + ... + n^k$ for $k = 2, 3, ...$ For this reason let us note the following:

1. The coefficients on the *dashed lines* remain constant.
2. The denominator of each coefficient on the *solid line* with arrows for each next row increases by 1. In other words, this means that this coefficient is equal to $\dfrac{1}{k+1}$.
3. *Small line segments with arrows* show how each next row can be obtained using the coefficients of the previous row. In other words, each time we multiply the coefficient of the previous row by a fraction with nominator equal to the value of k written in that row and with the denominator equal to the power of n written in that column.
4. If k is even, for example $k = 2$, then that row we start with B_2 and so on for the other even values of k, where $B_2, B_4, B_6...$ are *Bernoulli numbers*. The list of Bernoulli numbers is available in the literature and online, we would like to provide few of them $B_2 = \dfrac{1}{6}, B_4 = \dfrac{1}{-30}, B_6 = \dfrac{1}{42}, B_8 = \dfrac{1}{-30}, B_{10} = \dfrac{5}{66}, ...$

Example 1. Calculate the sum $1^3 + 2^3 + ... + n^3$.
Using this *power sums triangle* we deduce that

$$1^3 + 2^3 + ... + n^3 = 0 \cdot n + B_2 \cdot \dfrac{3}{2} \cdot n^2 + \dfrac{1}{2} \cdot n^3 + \dfrac{1}{4} \cdot n^4.$$

Hence, we obtain that
$$1^3 + 2^3 + ... + n^3 = \frac{1}{6} \cdot \frac{3}{2} \cdot n^2 + \frac{1}{2} \cdot n^3 + \frac{1}{4} \cdot n^4.$$
Thus, it follows that
$$1^3 + 2^3 + ... + n^3 = \frac{n^2}{4} + \frac{n^3}{2} + \frac{n^4}{4} = \left(\frac{n(n+1)}{2}\right)^2.$$

Example 2. Calculate the sum $1^6 + 2^6 + ... + n^6$.
Using this *power sums triangle* we deduce that
$$1^6 + 2^6 + ... + n^6 = B_6 \cdot n + 0 \cdot n^2 + B_4 \cdot \frac{5}{2} \cdot \frac{6}{3} \cdot n^3 + 0 \cdot n^4 + B_2 \cdot \frac{3}{2} \cdot \frac{4}{3} \cdot \frac{5}{4} \cdot \frac{6}{5} \cdot n^5 + \frac{1}{2} \cdot n^6 + \frac{1}{7} \cdot n^7.$$
Hence, we have that
$$1^6 + 2^6 + ... + n^6 = B_6 \cdot n + B_4 \cdot \frac{5}{2} \cdot \frac{6}{3} \cdot n^3 + B_2 \cdot \frac{6}{2} \cdot n^5 + \frac{1}{2} \cdot n^6 + \frac{1}{7} \cdot n^7.$$
Thus, it follows that
$$1^6 + 2^6 + ... + n^6 = \frac{1}{42} \cdot n - \frac{1}{30} \cdot \frac{5}{2} \cdot \frac{6}{3} \cdot n^3 + \frac{1}{6} \cdot \frac{6}{2} \cdot n^5 + \frac{1}{2} \cdot n^6 + \frac{1}{7} \cdot n^7.$$
Therefore, we obtain that
$$1^6 + 2^6 + ... + n^6 = \frac{n}{42} - \frac{n^3}{6} + \frac{n^5}{2} + \frac{n^6}{2} + \frac{n^7}{7}.$$

Remark. Note that in the literature a corresponding formula for the sum of 6th powers can be found, but for students it is almost impossible to memorize these formulas for the sum of any power, including the formula for the sum of 6th power. A straightforward verification shows that, the expression we have obtained is the same as the below mentioned formula that can be found in the literature, nevertheless our approach (*power sums triangle*) is self-constructive and can be easily applied to obtain a formula for the sum of any power.
$$1^6 + 2^6 + ... + n^6 = \frac{1}{42} n(n+1)(2n+1)(3n^4 + 6n^3 - 3n + 1).$$

1.1.9 Cartesian coordinate system, some important formulas.

Cartesian coordinate system. Cartesian coordinate system specifies each point uniquely in a plane by a set of numerical coordinates. Cartesian coordinate system was invented in 17th century by a prominent French mathematician *René Descartes*, his latinized name was *Cartesius*. His work revolutionized mathematics by providing link between *geometry* and *algebra*.

The distance formula in one dimension. The distance between two points $P_1 = x_1$ and $P_2 = x_2$ lying on the real number line can be found using the following formula.
$$d(P_1, P_2) = |x_1 - x_2|.$$

The distance formula in two dimensions. The distance between two points $P_1 = (x_1, y_1)$ and $P_2 = (x_2, y_2)$ of two dimensional xy–plane can be found using the following formula.
$$d(P_1, P_2) = \sqrt{(x_1 - x_2)^2 + (y_1 - y_2)^2}.$$

The distance formula in three dimensions. The distance between two points $P_1 = (x_1, y_1, z_1)$ and $P_2 = (x_2, y_2, z_2)$ of three dimensional xyz–space can be found using the following formula.
$$d(P_1, P_2) = \sqrt{(x_1 - x_2)^2 + (y_1 - y_2)^2 + (z_1 - z_2)^2}.$$

The distance formula in n−dimensional space. The distance between two points $P = (p_1, p_2, ..., p_n)$ and $Q = (q_1, q_2, ..., q_n)$ of n−dimensional space can be found using the following formula.
$$d(P, Q) = \sqrt{(q_1 - p_1)^2 + (q_2 - p_2)^2 + ... + (q_n - p_n)^2}.$$

The midpoint formula in n−dimensional space. The midpoint of a line segment in n−dimensional space with endpoints $P_1 = (x_1, x_2, ..., x_n)$ and $P_2 = (y_1, y_2, ..., y_n)$ is given by $\dfrac{P_1 + P_2}{2}$. In other words, the ith coordinate of the midpoint is $\dfrac{x_i + y_i}{2}$, where $i = 1, 2, ..., n$.

Equation of the line passing through two points. The equation of the line passing through two points $P_1 = (x_1, y_1)$ and $P_2 = (x_2, y_2)$ of two dimensional xy−plane can be found using the following formula.
$$y - y_1 = \frac{y_2 - y_1}{x_2 - x_1} \cdot (x - x_1).$$

Remark. The coefficient $m = \dfrac{y_2 - y_1}{x_2 - x_1}$ is called the slope of a staight line and represents the ratio of the "vertical change" to the "horizontal change".

Parallel lines. Two lines are parallel if they have the same slope.
That is, if the equation of the first line is $y = m_1 x + b_1$ and the equation of the second line is $y = m_2 x + b_2$, then these two lines are parallel if $m_1 = m_2$ and $b_1 \neq b_2$.

Perpendicular lines. Two lines are perpendicular, if the product of their slopes is equal to -1.
That is, if the equation of the first line is $y = m_1 x + b_1$ and the equation of the second line is $y = m_2 x + b_2$, then these two lines are perpendicular if $m_1 \cdot m_2 = -1$.

1.2 Geometry (AMC 10): the most useful formulas and theorems

1.2.1 Formulas for plane shapes.

Formula of the inradius of a right triangle. For inradius r of a right triangle with legs a, b and hypotenuse c, we have that
$$r = \frac{a+b-c}{2}.$$

Pythagorean theorem. For a right triangle with legs a, b and hypotenuse c, we have that
$$a^2 + b^2 = c^2.$$

Law of cosines. In a triangle ABC let a, b, c be the lengths of sides opposite to $\angle A, \angle B, \angle C$, respectively, then
$$\begin{cases} a^2 = b^2 + c^2 - 2bc \cdot \cos \angle A, \\ b^2 = a^2 + c^2 - 2ac \cdot \cos \angle B, \\ c^2 = a^2 + b^2 - 2ab \cdot \cos \angle C. \end{cases}$$

Remark. Note that the Pythagorean theorem is a special case of the law of cosines, because when $\angle C = 90°$ then $\cos \angle C = \cos 90° = 0$.

Law of sines. In a triangle ABC let a, b, c be the lengths of sides opposite to the angles A, B, C, respectively, then
$$\frac{a}{\sin A} = \frac{b}{\sin B} = \frac{c}{\sin C}.$$

Area of a triangle.
$$S = \frac{a \cdot h_a}{2} = \frac{rp}{2} = \frac{1}{2} ab \cdot \sin \gamma = \frac{abc}{4R},$$
where S is the area of the triangle, a, b, c are the side lengths, h_a is the length of the altitude to side a, p is the perimeter, r is the inradius, R is the circumradius, γ is the angle between the sides a and b.

Area of an equilateral triangle.
$$S = \frac{\sqrt{3}}{4} a^2,$$
where a is the side length.

Area of a rectangle and a square. Area of a rectangle is given by the following formula $S = a \cdot b$, where a, b are the side lengths of the rectangle. Note that in the case of the square $a = b$, therefore area of a sqaure is given by the following formula $S = a^2$.

Area of a trapezoid.
$$S = \frac{(b_1 + b_2) \cdot h}{2} = m \cdot h,$$
where b_1 and b_2 are the heights of the trapezoid, h is the height, m is the midsegment of the trapezoid (midsegment or midline is the segment connecting the midpoints of the two non-parallel sides).

Area of a quadrilateral (general formula). In the subsection *Bretschneider's formula, diagonals and sides area formula of a quadrilateral* we provide useful formulas to find the area of a quadrilateral, these formulas work for any quadrilateral. Unlike other general formulas of a quadrilateral, there is no need to calculate angles or other distances in the quadrilateral, it is sufficient to have only the lengths of the sides and the lengths of the diagonals. We also provide *Bretschneider's formula* to find the area of a quadrilateral, this formula works for any quadrilateral using only the side lengths and the sum of two opposite angles.

Area of a circle, circumference, area of a sector. Area of a circle is given by the following formula $S = \pi \cdot r^2$, circumference of a circle $C = 2\pi \cdot r$, where r is the radius of the circle.

Area of a regular hexagon.
$$S = \frac{3\sqrt{3}a^2}{2},$$
where a is the side length.

Area of a regular polygon.
$$S = \frac{na^2}{4\tan\dfrac{180°}{n}} = \frac{hp}{2},$$
where n is the number of sides, a is the side length, h is the apothem, p is the perimeter.

1.2.2 Tangential, cyclic, bicentric and extangential quadrilaterals, Pitot's theorem.

Tangential quadrilateral (or circumscribed quadrilateral). A tangential quadrilateral (or circumscribed quadrilateral) is a convex quadrilateral whose all four sides are tangent to a single circle within the quadrilateral.

Remark (incircle and inradius). This circle is called *incircle* and its radius is called *inradius* of the quadrilateral.

Pitot's theorem. A quadrilateral is tangential if and only if the sums of lengths of opposite sides are the same.

The direct statement of this theorem was proved by a French scientist *Henri Pitot*, the converse statement was proved later on by a Swiss mathematician *Jakob Steiner*.

Tangential quadrilateral theorem. A convex quadrilateral is tangential if and only if the angle bisectors of its four angles intersect at one point (are concurrent). Moreover, the intersection point is the center of the incircle (incenter).

Tangent lengths. Eight *tangent lengths* of a tangential quadrilateral are the line segments from a vertex to the points of tangency of its incircle and its sides.

Tangent lenghts theorem. In a tangential quadrilateral two tangent lengths corresponding to each vertex are congruent to each other (are equal in lenght).

Remark. This theorem is equivalent to the following statement: two tangents to a circle from a given point are equal in lenght to where they touch the circle (points of tangency).

The area S of a tangential quadrilateral can be given by different formulas, the simplest one is the following formula:
$$S = r \cdot s,$$
where s is the semiperimeter of the quadrilateral and r is the inradius.

Remark. This formula holds true for any tangential polygon.

Note that, if we denote the opposite sides of the tangential quadrilateral by a, c and b, d, then from the definition of a semi-perimeter and from Pitot's theorem we have that
$$s = \frac{a+b+c+d}{2} = a+c = b+d.$$

Remark (alternative formula of the area of a tangential quadrilateral.) Above mentioned formula is not the only formula of the area of a tangential quadrilateral. Its area can be given by different formulas, in the subsection *Bretschneider's formula, diagonals and sides area formula of a quadrilateral* we provide another useful formula of the area of a tangential quadrilateral, where the area can be calculated using only the lenghts of its diagonals and sides.

Cyclic quadrilateral (or inscribed quadrilateral). A cyclic quadrilateral (or inscribed quadrilateral) is a quadrilateral whose all vertices lie on one circle.

Remark (circumcircle and circumradius). This circle is called *circumcircle* and its radius is called *circumradius* of the quadrilateral.

Cyclic quadrilateral theorem 1. A convex quadrilateral is cyclic if and only if its opposite angles are supplementary (they add up to $180°$).

The direct statements of theorem 1, theorem 2 and theorem 3 were given by a prominent mathematician of Greco-Roman antiquity *Euclid* (Eukleides of Alexandria, reffered as one of the founders of Geometry).

Consequence. A convex quadrilateral is cyclic if and only if an exterior angle is equal to the opposite interior angle.

Cyclic quadrilateral theorem 2. A convex quadrilateral is cyclic if and only if the angle between a side and a diagonal is equal to the angle between the opposite side and the other diagonal.

As we have already mentioned the direct statement of this theorem was given by *Euclid*, the converse statement was proved later on by a French mathematician *Jacques Hadamard*.

Cyclic quadrilateral theorem 3. Let $ABCD$ be a convex quadrilateral, such that lines AB and CD intersect at point E, then $ABCD$ is cyclic if and only if

$$AE \cdot EB = DE \cdot EC.$$

Cyclic quadrilateral theorem 4. A convex quadrilateral is cyclic if and only if four perpendicular bisectors to its sides intersect at the same point (are concurrent). Note that this intersection point is the circumcenter of the quadrilateral.

The area of a cyclic quadrilateral can be given by different formulas, the simplest one is Brahmagupha's formula (see the next subsection).

Bicentric quadrilateral. A bicentric quadrilateral is a convex quadrilateral that is both tangential and cyclic (has simultaneously an incircle and a circumcircle).

Extangential quadrilateral. Extangential quadrilateral is a convex quadrilateral whose extensions of all four sides are tangent to a circle outside the quadrilateral.

Remark (excircle and exradius.) This circle is called *excircle* and its radius is called *exradius* of the quadrilateral.

Extangential quadrilateral theorem. A convex quadrilateral is extangential if and only if the sum of its some two adjacent sides is equal to the sum of the other two sides.

This theorem was proved by a Swiss mathematician *Jakob Steiner*. Note that there are some other characterization theorems for extangential quadrilaterals (related to its angles), but we do not cover them in this book, as it is outside the scope of AMC.

1.2.3 Heron's formula and Brahmagupha's formula.

Heron's formula. *Heron's formula* is named after a well-known mathematician and experimenter of Greco-Roman antiquity *Hero of Alexandria*, also called *Heron of Alexandria*. This formula is applied to find the area S of any triangle

$$S = \sqrt{s(s-a)(s-b)(s-c)},$$

where a, b, c are the side lengths of given triangle and s is the semi-perimeter of the triangle, that is

$$s = \frac{a+b+c}{2}.$$

Brahmagupta's formula. *Brahmagupta's formula* is named after an Indian mathematician and astronomer *Brahmagupta*. This formula is applied to find the area S of any *cyclic* quadrilateral

$$S = \sqrt{(s-a)(s-b)(s-c)(s-d)},$$

where a, b, c, d are the side lengths of given cyclic quadrilateral and s is the semi-perimeter of the quadrilateral, that is

$$s = \frac{a+b+c+d}{2}.$$

1.2.4 Ceva's theorem, Menelaus' theorem, Stewart's theorem, Ptolemy's theorem.

Ceva's theorem. Given a triangle ABC. Let points D, E, F be on sides BC, AC, AB, respectively. Line segments AD, BE, CF intersect at one point (are concurrent), if and only if

$$\frac{AF}{FB} \cdot \frac{BD}{DC} \cdot \frac{CE}{EA} = 1.$$

Ceva's theorem is named after an Italian mathematician *Giovanni Ceva*. Nevertheless, it was proved much earlier by other authors too.

Remark (cevians). Line segments AD, BE, CF are known as *cevians*.

Menelaus' theorem. Given a triangle ABC. Let D, E be given points on sides BC, AC, respectively, and F be a given point on line AB (outside side AB). Points D, E, F lie on the same line (are collinear), if and only if

$$\frac{FA}{FB} \cdot \frac{DB}{DC} \cdot \frac{EC}{EA} = 1.$$

Menelaus' theorem is named after a mathematician of Greco-Roman antiquity *Menelaus of Alexandria*.

Stewart's theorem. Given a triangle ABC. Let D be a given point on side BC, then

$$AD^2 = \frac{AB^2 \cdot CD + AC^2 \cdot BD}{BC} - BD \cdot CD.$$

Stewart's theorem is named after a Scottish mathematician *Mattew Stewart* and it is used to express the length of a cevian in a triangle by the lengths of the sides.

Ptolemy's theorem. Quadrilateral $ABCD$ is a cyclic quadrilateral if and only if

$$AB \cdot CD + BC \cdot AD = AC \cdot BD.$$

Ptolemy's theorem is named after a mathematician of Greco-Roman antiquity *Claudius Ptolemaeus*, also known as *Ptolemy (of Alexandria)*.

Remark (Ptolemy's inequality). Let $ABCD$ be a quadrilateral, then

$$AB \cdot CD + BC \cdot AD \geq AC \cdot BD,$$

where the equality holds true if and only if $ABCD$ is a cyclic quadrilateral.

1.2.5 Bretschneider's formula, *diagonals and sides* area formula of a quadrilateral.

In this subsection we provide area formulas for a quadrilateral. The first formula we call *diagonals and sides* area formula of a quadrilateral, as for finding the area we use the lengths of its diagonals and sides.

***Diagonals and sides* area formula of a quadrilateral.** The area of any convex quadrilateral is given by the following general formula

$$S = \frac{1}{4}\sqrt{4e^2f^2 - (b^2 + d^2 - a^2 - c^2)^2},$$

where S is the area, e, f are the lenghts of the diagonals and a, b, c, d are the lengths of the sides.

Alternative formulation of the (*diagonals and sides*) area formula of a quadrilateral.

$$S = \sqrt{(s-a)(s-b)(s-c)(s-d) - \frac{1}{4}(ac+bd+ef)(ac+bd-ef)},$$

where S is the area, s is the semi-perimeter and a, b, c, d are the lengths of the sides of the quadrilateral.

Another alternative but equivalent formula written in trigonometric form is the following one.
Bretschneider's formula.

$$S = \sqrt{(s-a)(s-b)(s-c)(s-d) - abcd\cos^2\left(\frac{\alpha+\gamma}{2}\right)},$$

where S is the area, s is the semi-perimeter, a, b, c, d are the lengths of the sides and α, γ are two opposite angles of the quadrilateral.

This result was published by a German mathematician *C. A. Bretschneider*. Besides Bretschneider this result was published by other authors too. The first two results were published by several authors also, in particular by a British mathematician *E.W. Hobson* and the alternative but equivalent formulation by the American mathematician *J.L. Coolidge*.

Cyclic quadrilateral area formula. Note that for a cyclic quadrilateral using either Bretschneider's formula with the property $\alpha + \gamma = 180°$ or the alternative formulation of *diagonals and sides* area formula with Ptolemy's theorem $ac + bd = ef$ one can easily deduce Brahmagupta's formula.

$$S = \sqrt{(s-a)(s-b)(s-c)(s-d)}.$$

Using the *diagonals and sides* area formula of a quadrilateral and Pitot's theorem ($a + c = b + d$), we have deduced the following formula of the area for any tangential quadrilateral.

Tangential quadrilateral area formula. The area of a tangential quadrilateral can given by the following formula.

$$S = \frac{1}{2}\sqrt{e^2 f^2 - (ac-bd)^2}.$$

1.2.6 Parameshvara's formula for circumradius.

Parameshvara's formula The circumradius R of a cyclic quadrilateral can be given by the following formula

$$R = \frac{1}{4}\sqrt{\frac{(ab+cd)(ac+bd)(ad+bc)}{(p-a)(p-b)(p-c)(p-d)}}$$

where a, b, c, d are the lengths of the sides and p is the semi-perimeter of given cyclic quadrilateral. Note that, using Brahmagupta's formula we can rewrite Parameshvara's formula in the following form

$$R = \frac{\sqrt{(ab+cd)(ac+bd)(ad+bc)}}{4 \cdot S},$$

where S is the area of given cyclic quadrilateral.

Parameshvara's formula is named after an Indian mathematician *Vatasseri Parameshvara Nambudiri*.

1.2.7 Formulas for volume and surface areas of three-dimensional shapes.

Volume and surface area of a cube.
$$V = a^3,$$
$$S = 6a^2,$$
where a is the side length of the cube.

Volume and surface area of a cone.
$$V = \frac{\pi r^2 h}{3},$$
$$S = \pi r^2 + \pi r l,$$
where r is the radius of the circular base, h is the height of the cone, l is the slant height.

Volume and surface area of a tetrahedron/pyramid.
$$V = \frac{Ah}{3},$$
where A is the area of the base, h is the height of the tetrahedron/pyramid.

Regular pyramid. A regular pyramid is a pyramid whose base is a regular polygon and whose lateral edges are all equal in length.
In the case of a regular pyramid its total surface area can be given by the following formula.
$$S = \frac{a \cdot l}{2} \cdot n + A = \frac{a \cdot l}{2} \cdot n + \frac{n \cdot a^2}{4 \tan \frac{180°}{n}},$$
where a is the lenght of the side, l is the slant height, n is the number of sides of the base (base is a regular polygon), A is the area of the base.

Volume and surface area of a cylinder.
$$V = \pi r^2 h,$$
$$S = 2\pi r^2 + 2\pi r h,$$
where r is the radius of the circular base, h is the height.

Volume and surface area of a rectangular prism.
$$V = lwh,$$
$$S = 2(lw + lh + wh),$$
where l is the length, w is the width, h is the height.

1.2.8 Trigonometric identities.

Some problems in the AMC 10, AMC 12 and many other math competitions may be solved using trigonometry. Mostly either by application of basic trigonometric identities or using the values of sines and cosines of common angles, such as

$$\sin 30° = \sin \frac{\pi}{6} = \cos 60° = \frac{1}{2},$$

$$\sin 45° = \sin \frac{\pi}{4} = \cos 45° = \frac{\sqrt{2}}{2}.$$

For more values of sines and cosines of common angles see *trigonometric unit circle* provided below.

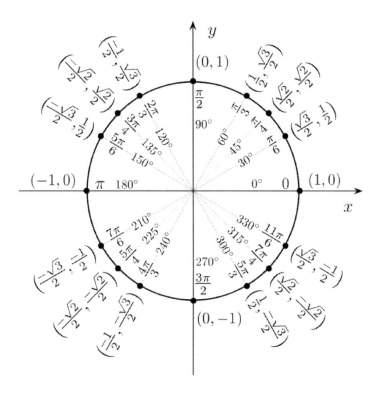

Quotient identities.

$$\tan\alpha = \frac{\sin\alpha}{\cos\alpha}.$$

$$\cot\alpha = \frac{\cos\alpha}{\sin\alpha}.$$

Reciprocal identities.

$$\cot\alpha = \frac{1}{\tan\alpha}.$$

$$\csc\alpha = \frac{1}{\sin\alpha}.$$

$$\sec\alpha = \frac{1}{\cos\alpha}.$$

Pythagorean identities.

$$\sin^2\alpha + \cos^2\alpha = 1.$$

$$\tan^2\alpha + 1 = \frac{1}{\cos^2\alpha}.$$

$$\cot^2\alpha + 1 = \frac{1}{\sin^2\alpha}.$$

Sum-to-product identities.

$$\sin\alpha + \sin\beta = 2\sin\frac{\alpha+\beta}{2}\cos\frac{\alpha-\beta}{2}.$$

$$\sin\alpha - \sin\beta = 2\sin\frac{\alpha-\beta}{2}\cos\frac{\alpha+\beta}{2}.$$

$$\cos\alpha + \cos\beta = 2\cos\frac{\alpha+\beta}{2}\cos\frac{\alpha-\beta}{2}.$$

$$\cos\alpha - \cos\beta = -2\sin\frac{\alpha+\beta}{2}\sin\frac{\alpha-\beta}{2}.$$

Product-to-sum identities.

$$\sin\alpha\sin\beta = \frac{1}{2}(\cos(\alpha-\beta) - \cos(\alpha+\beta)).$$

$$\cos\alpha\cos\beta = \frac{1}{2}(\cos(\alpha-\beta) + \cos(\alpha+\beta)).$$

$$\sin\alpha\cos\beta = \frac{1}{2}(\sin(\alpha-\beta) + \sin(\alpha+\beta)).$$

Sum identities.

$$\sin(\alpha+\beta) = \sin\alpha\cos\beta + \cos\alpha\sin\beta.$$

$$\cos(\alpha+\beta) = \cos\alpha\cos\beta - \sin\alpha\sin\beta.$$

$$\tan(\alpha+\beta) = \frac{\tan\alpha + \tan\beta}{1 - \tan\alpha\tan\beta}.$$

Difference identities.

$$\sin(\alpha-\beta) = \sin\alpha\cos\beta - \cos\alpha\sin\beta.$$

$$\cos(\alpha-\beta) = \cos\alpha\cos\beta + \sin\alpha\sin\beta.$$

$$\tan(\alpha-\beta) = \frac{\tan\alpha - \tan\beta}{1 + \tan\alpha\tan\beta}.$$

Cofunction identities.

$$\sin\left(\frac{\pi}{2} - \alpha\right) = \cos\alpha.$$

$$\cos\left(\frac{\pi}{2} - \alpha\right) = \sin\alpha.$$

$$\tan\left(\frac{\pi}{2} - \alpha\right) = \cot\alpha.$$

$$\cot\left(\frac{\pi}{2} - \alpha\right) = \tan\alpha.$$

$$\csc\left(\frac{\pi}{2} - \alpha\right) = \sec\alpha.$$

$$\sec\left(\frac{\pi}{2} - \alpha\right) = \csc\alpha.$$

Negative angle (even/odd) identities.

$$\sin(-\alpha) = -\sin\alpha.$$

$$\cos(-\alpha) = \cos\alpha.$$

$$\tan(-\alpha) = -\tan\alpha.$$

$$\cot(-\alpha) = -\cot\alpha.$$

$$\csc(-\alpha) = -\csc\alpha.$$

$$\sec(-\alpha) = \sec\alpha.$$

Double-angle identities.

$$\sin 2\alpha = 2\sin\alpha\cos\alpha.$$

$$\cos 2\alpha = \cos^2\alpha - \sin^2\alpha = 2\cos^2\alpha - 1 = 1 - 2\sin^2\alpha.$$

$$\tan 2\alpha = \frac{2\tan\alpha}{1 - \tan^2\alpha}.$$

Half-angle identities.

$$\sin\frac{\alpha}{2} = \pm\sqrt{\frac{1-\cos\alpha}{2}}.$$

$$\cos\frac{\alpha}{2} = \pm\sqrt{\frac{1+\cos\alpha}{2}}.$$

$$\tan\frac{\alpha}{2} = \frac{\sin\alpha}{1+\cos\alpha} = \frac{1-\cos\alpha}{\sin\alpha}.$$

Triple angle identities.

$$\sin 3\alpha = 3\sin\alpha - 4\sin^3\alpha.$$

$$\cos 3\alpha = 4\cos^3\alpha - 3\cos\alpha.$$

$$\tan 3\alpha = \frac{3\tan\alpha - \tan^3\alpha}{1 - 3\tan^2\alpha}.$$

Symmetry identities.

$$\sin(\pi - \alpha) = \sin\alpha.$$
$$\cos(\pi - \alpha) = -\cos\alpha.$$
$$\tan(\pi - \alpha) = -\tan\alpha.$$
$$\cot(\pi - \alpha) = -\cot\alpha.$$
$$\csc(\pi - \alpha) = \csc\alpha.$$
$$\sec(\pi - \alpha) = -\sec\alpha.$$
$$\sin(\pi + \alpha) = -\sin\alpha.$$
$$\cos(\pi + \alpha) = -\cos\alpha.$$
$$\tan(\pi + \alpha) = \tan\alpha.$$
$$\cot(\pi + \alpha) = \cot\alpha.$$
$$\csc(\pi + \alpha) = -\csc\alpha.$$
$$\sec(\pi + \alpha) = -\sec\alpha.$$

1.2.9 Complex numbers, de Moivre's formula, Euler's formula.

Complex numbers. A complex number is a number that can be expressed in the form $a + bi$, where a, b are real numbers and i is an indeterminate satisfying $i^2 = -1$. As no real number satisfies this equation, i is called an *imaginary number*, a is called the *real part* and b is called the *imaginary part*.

Remark. The real part of a complex number z is denoted by $Re(z)$, the imaginary part of a complex number z is denoted by $Im(z)$.

Example. Consider complex number $5 + 3i$, then $Re(5 + 3i) = 5$ and $Im(5 + 3i) = 3$.

Absolute value (modulus or magnitude) of a complex number. The *absolute value* (modulus or magnitude) of a complex number $z = a + bi$ is denoted by $|z|$ and is equal to:

$$|z| = \sqrt{a^2 + b^2}.$$

Conjugate of a complex number. The complex *conjugate* of complex number $z = a + bi$ is given by $a - bi$ and is denoted by $\overline{z}$.

Remark. Note that

$$z \cdot \overline{z} = (a + bi)(a - bi) = a^2 + b^2 = |z|^2 = |\overline{z}|^2.$$

De Moivre's formula. For any real number x and integer n it holds that

$$(\cos x + i\sin x)^n = \cos(nx) + i\sin(nx),$$

where i is the imaginary number.

This formula is named after a French mathematician *Abraham de Moivre*. De Moivre's formula can be considered as a consequence of the below mentioned *Euler's formula*.

Euler's formula. Euler's formula establishes the following relationship between the trigonometric functions and the complex exponential function:

$$e^{ix} = \cos x + i\sin x,$$

where e is called *Euler's number* after a Swiss mathematician *Leonhard Euler* and i the imaginary number. Mathematical constant e is approximately equal to 2.71828 and is the base of the natural logarithm.

Remark. The expression $\cos x + i\sin x$ sometimes is denoted as $cis(x)$.

1.3 Number theory (AMC 10): the most useful formulas and theorems

1.3.1 Unique-prime-factorization theorem (fundamental theorem of arithmetic).

Unique-prime-factorization theorem (fundamental theorem of arithmetic). Every integer greater than 1 is either is prime number or can be represented as the product of prime numbers. Moreover, this representation is unique, up to the order of the factors.

Mathematically, this statement can be rewritten in the following way.

Unique-prime-factorization theorem (fundamental theorem of arithmetic). Every integer n greater than 1 can be represented as

$$n = p_1^{\alpha_1} \cdot p_2^{\alpha_2} \cdot \ldots \cdot p_k^{\alpha_k},$$

where $p_1, p_2, ..., p_k$ are distinct primes and $\alpha_1, \alpha_2, ..., \alpha_k$ are positive integers.

Remark. This representation is called the **canonical representation** of n or **standard form** of n.

1.3.2 Number of divisors of a composite number, sum and product of divisors.

Number of divisors of a composite number n. If the prime-factorization of a composite number n is $n = p_1^{\alpha_1} \cdot p_2^{\alpha_2} \cdot \ldots \cdot p_k^{\alpha_k}$, then the number of divisors of n (denoted by $d(n)$) is equal to

$$d(n) = (\alpha_1 + 1)(\alpha_2 + 1)...(\alpha_k + 1).$$

Sum of divisors of a composite number. If the prime-factorization of a composite number n is $n = p_1^{\alpha_1} \cdot p_2^{\alpha_2} \cdot \ldots \cdot p_k^{\alpha_k}$, then the sum of divisors of n (denoted by $\tau(n)$) is equal to

$$\tau(n) = \frac{p_1^{\alpha_1+1} - 1}{p_1 - 1} \cdot \frac{p_2^{\alpha_2+1} - 1}{p_2 - 1} \cdot \ldots \cdot \frac{p_k^{\alpha_k+1} - 1}{p_k - 1}.$$

Product of divisors of a composite number. If the prime-factorization of a composite number n is $n = p_1^{\alpha_1} \cdot p_2^{\alpha_2} \cdot \ldots \cdot p_k^{\alpha_k}$, then the product of divisors of n (denoted by $\pi(n)$) is equal to

$$\pi(n) = n^{\frac{d(n)}{2}},$$

where as mentioned above $d(n)$ is the number of divisors of n.

1.3.3 One useful lemma.

The following lemma can be applied as a very useful technique to treat non-standard *number theory* problems.

One useful lemma. Let n be a positive integer. Let $d_1, d_2, ..., d_k$ be all positive integer divisors of n, such that $1 = d_1 < d_2 < ... < d_k = n$, then

$$d_1 = \frac{n}{d_k}, d_2 = \frac{n}{d_{k-1}}, ..., d_k = \frac{n}{d_1}.$$

1.3.4 Modular arithmetic and congruence relation.

Congruence relation. For a positive integer n integers a and b are called *congruent modulo n* if $n \mid a-b$.
Remark. This notation means that n divides $a - b$. In other words it means that $a - b$ is divisible by n, that is, if there exits an integer k such that $a - b = k \cdot n$.
Congruence modulo n is denoted in the following way.

$$a \equiv b \ (mod \ n).$$

1.3.5 Fermat's little theorem and Wilson's theorem.

Fermat's little theorem. If p is prime number, then for any integer a, we have that

$$p \mid a^p - a.$$

Remark. Using *modular arithmetic* and *congruence relation* Fermat's little theorem can be rewritten in the following way

$$a^p \equiv a \ (mod \ p).$$

Fermat's little theorem (alternative formulation). If p is prime number, then for any integer a, we have that

$$p \mid a^{p-1} - 1.$$

Remark. Using *modular arithmetic* and *congruence relation* alternative formulation of Fermat's little theorem can be rewritten in the following way

$$a^{p-1} \equiv 1 \ (mod \ p).$$

Fermat's little theorem is named after a prominent French mathematician and lawyer *Pierre de Fermat*.
Wilson's theorem. Integer number n greater than 1, is prime number if and only if

$$n \mid (n-1)! + 1.$$

Remark. Using *modular arithmetic* and *congruence relation* Wilson's theorem can be rewritten in the following way

$$(n-1)! \equiv -1 \ (mod \ n).$$

Wilson's theorem is named after an English mathematician *John Wilson*. It is funny, but Wilson was neither the first person to state this theorem nor the first person to prove it. A prominent French mathematician of Italian descent *Joseph-Louis Lagrange* (born *Giuseppe Luigi Lagrangia*) gave the first published proof of this theorem. Several other mathematicians have stated this result earlier, without providing a published proof.

1.4 Combinatorics and probability (AMC 10): the most useful formulas and theorems

1.4.1 Rule of sum and rule of product.

In combinatorics, the rule of sum (or addition principle) and the rule of product (or multiplication principle) are basic counting principles.

The rule of sum (or addition principle). Let n and m be nonnegative integers. If there are n choices for one action and m choices for another action and these two actions cannot be done simultaneously, then there are $n + m$ ways to perform one of these actions.

The rule of product (or multiplication principle). Let n and m be nonnegative integers. If there are n choices for one action and m choices for another action, then there are $n \cdot m$ ways to perform both of these nonnegative actions.

1.4.2 Permutations.

Permutation. A permutation of a set of objects is an ordering (rearrangement) of these objects.

Permutation with repetition (or r–tuple). Let r and n be nonnegative integers, such that $n \geq r$. A permutation with repetition (or r–tuple) is an ordered selection of r elements from a set of n elements, where repetition is allowed.

Theorem (the number of permutations with repetition). Let r and n be nonnegative integers, such that $n \geq r$. The number of permutations with repetition for selection r elements from a set of n elements where repetition is allowed is equal to n^r.

Permutation without repetition. Let k and n be nonnegative integers, such that $n \geq k$. A permutation without repetition is an ordered selection of k elements from a set of n elements, where repetition is not allowed.

Theorem (the number of permutations without repetition). Let k and n be nonnegative integers, such that $n \geq k$. The number of permutations without repetition of obtaining an ordered subset of k elements from a set of n elements is denoted by nP_k.

$$nP_k = \frac{n!}{(n-k)!} = n(n-1) \cdot ... \cdot (n-(k-1)),$$

where $n! = 1 \cdot 2 \cdot ... \cdot n$.

Corollary. Let n be a positive integer. The number of permutations without repetition of obtaining an ordered subset of n elements from a set of n elements is denoted by P_n and

$$P_n = n!.$$

1.4.3 Combinations.

Combination. Let k and n be nonnegative integers, such that $n \geq k$. The number of ways of obtaining an unordered subset of k elements from a set of n elements is called combination and is denoted by $\binom{n}{k}$ or C_n^k or nC_k.

Remark 1 (binomial coefficent). The symbol $\binom{n}{k}$ is called the binomial coefficient.

Remark 2. The symbols $\binom{n}{k}$ or nC_k or C_n^k are read as n *choose* k, meaning that there are $\binom{n}{k}$ ways to choose an unordered subset of k elements from a fixed set of n elements.

Remark 3. One can say that a permutation is an "ordered combination."

Theorem (the number of combinations.) Let k and n be nonnegative integers, such that $n \geq k$. The number of combinations of obtaining an unordered subset of k elements from a set of n elements can be found by the following formula.

$$\binom{n}{k} = \frac{n!}{k!(n-k)!}.$$

1.4.4 Stars and bars technique (integer equations).

In combinatorics very often arise problems such as counting the number of ways to group identical objects, for example placing indistinguishable balls into ennumerated urns, or finding the number of nonnegative integer solutions an equation has. Stars and bars technique is a short and elegant way to treat such problems. We provide the following example to explain how this technique works.

Example. Consider the following equation

$$a + b + c + d + e = 16,$$

where a, b, c, d, e are nonnegative integers. What is the total number of all nonnegative integer solutions of this equation?

Solution. Assume there are 20 places, where we place 16 stars and 4 bars (one per place).
The main idea is that any such arrangment represents a solution of given equation. For example

$$**\,|\,****\,|\,***\,|\,******\,|\,*$$

represents the solution

$$2 + 4 + 3 + 6 + 1 = 16.$$

Remark. Note that before the first bar and after the last bar it is possible to place 0 stars.
Therefore, given problem is equivalent to the following question: *In how many ways is it possible to put 16 stars and 4 bars in 20 places?*. It is equivalent to fixing 4 places out of 20 places and putting stars in the empty places (one star per place). This can be done by $\binom{20}{4}$ ways. Using the corresponding formula for the binomail coefficient, we obtain that

$$\binom{20}{4} = 4845.$$

1.4.5 Probability.

Probability is a numerical description of how likely an event is to occur (or how likely is that a proposition holds true).

Probability is a number from the interval $[0, 1]$, where 0 stands for impossibility and 1 stands for certainty.

Probability range. $0 \leq P(event) \leq 1$.

Theoretical probability expresses the likelihood that something will occur.
Theoretical probability is equal to the number of favorable outcomes divided by the total number of possible outcomes.

Theoretical probability formula.

$$P(event) = \frac{number\ of\ favorable\ outcomes}{number\ of\ possible\ outcomes}.$$

Complementary events. Complementary probability is the probability of a given event not occurring (or of a different event occurring that can only occur if the first event does not occur).

Complementary events formula. Let A be a given event, we denote by A^C its complementary event. Then, we have that

$$P(A^C) + P(A) = 1.$$

For example, the probability of a tossed coin landing on heads is $\frac{1}{2}$ and the complementary probability of the tossed coin not landing on heads (landing on tails) is

$$1 - \frac{1}{2} = \frac{1}{2}.$$

Independent events. Two events are called independent if the fact that one event occurs does not affect the probability of the other event occurring.

For example, if someone tosses a coin, then the possible result (head or tail) is not affected by previous tosses.

Independent events formula. Events A and B are independent if and only if

$$P(A \text{ and } B) = P(A \cap B) = P(A) \cdot P(B).$$

For example, if two coins are tossed the probability of both being heads is

$$\frac{1}{2} \cdot \frac{1}{2} = \frac{1}{4}.$$

Disjoint events. Two events are called disjoint if they never occur simultaneously. In the literature, they are also called mutually exclusive events.

Disjoint events formula. If A and B are disjoint events, then the probability that both of them occur simultaneously is:

$$P(A \cap B) = 0.$$

Rule of addition of probability. If A and B are two events in a probability experiment, then the probability that either one of them occurs is:

$$P(A \cup B) = P(A) + P(B) - P(A \cap B).$$

Obviously, if A and B are disjoint events, then we have that:

Rule of addition of probability for disjoint events.

$$P(A \cup B) = P(A) + P(B).$$

Chapter 2

AMC 10 type practice tests

2.1 AMC 10 type practice test 1

Problem 2.1. *A train left City A at 8 : 00 AM and arrived to City B after 45 minutes. It stopped in City B for 10 minutes and continued to City C. The train covered the distance between City B and City C in 75 minutes. At what time did the train arrive to City C?*

(A) 11:00 AM (B) 10:05 AM (C) 10:00 AM (D) 10:10 AM (E) 10:30 AM

Problem 2.2. *Let M and N be the midpoints of sides BC and CD of rectangle ABCD, respectively. The area of triangle AMN is what percent of the area of rectangle ABCD?*

(A) 25 (B) 37.5 (C) 30 (D) 24 (E) 50

Problem 2.3. *Let $C(n)$ be the sum of all distinct prime divisors of a positive integer n, where $n > 1$. What is $C(C(C(2008)))$?*

(A) 19 (B) 34 (C) 253 (D) 30 (E) 20

Problem 2.4. *Six painters working at the same constant rate, can completely paint an apartment in 8 hours. How many painters were working if it took 12 hours to paint that apartment?*

(A) 5 (B) 3 (C) 4 (D) 2 (E) 1

Problem 2.5. *What is the value of the expression* $\dfrac{1^2 + 1 \cdot 2 + 2^2}{1^3 \cdot 2^3} + \dfrac{2^2 + 2 \cdot 3 + 3^2}{2^3 \cdot 3^3} + ... + \dfrac{9^2 + 9 \cdot 10 + 10^2}{9^3 \cdot 10^3}$?

(A) 1 (B) 0.9 (C) 0.99 (D) 0.5 (E) 0.999

Problem 2.6. *The path from A to B consists of a 10 kilometers stretch of flat land, 12 kilometers uphill and 16 kilometers downhill. A car covered the respective stretches at speeds of 80 kmh, 48 kmh, and 90 kmh. What was car's average speed in kmh (to the nearest integer number) while traveling from A to B?*

(A) 68 (B) 69 (C) 70 (D) 72 (E) 60

Problem 2.7. *What is the value of the expression* $\dfrac{(2^{2008})^2 + 2^{2007}}{(2^{2007})^2 + 2^{2005}}$?

(A) 2^{2007} (B) 2^{2008} (C) 9 (D) 4 (E) 5

Problem 2.8. *A shop bought a coat for $500. The shop inteded to sell the coat for some price, but sold it 5% more than the intended sales price. Given that for that deal the shop generated a total profit of $67. What was the intended sales price of the coat?*

(A) $560   (B) $540   (C) $550   (D) $530   (E) $520

Problem 2.9. *For how many positive values of x is $\dfrac{20}{x+1} + \dfrac{1}{3(x+1)}$ a positive integer?*

(A) 0   (B) 10   (C) 19   (D) 20   (E) 25

Problem 2.10. *Let S_1 be a square with a side length 7. The vertices of square S_2 are located on the sides of S_1. The sides of square S_3 are parallel to the sides of S_1, and the vertices of square S_3 are located on the sides of square S_2. Given that the side length of square S_3 is equal to 4. What is the value of the side length of square S_2?*

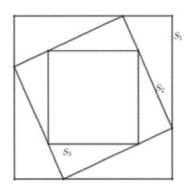

(A) $\sqrt{13}$ (B) $2\sqrt{7}$ (C) 6   (D) 5   (E) $4\sqrt{2}$

Problem 2.11. *A fisherman makes a round-trip and takes his boat 10 miles into a lake from its shore and back (covering the same distance). The average speed of the boat is 2.5 miles per hour. The fisherman catches (in average) 0.5 pounds of fish per hour. How many pounds of fish does the fisherman catch per one round-trip?*

(A) 4   (B) 3   (C) 2   (D) 5   (E) 6

Problem 2.12. *In one school there are 25% less 11th graders than 10th graders, and 20% more 11th graders than 12th graders. The total number of students in 10th, 11th, and 12th grades in that school is equal to 190. How many 10th graders are there in that school?*

(A) 60   (B) 50   (C) 40   (D) 100   (E) 80

Problem 2.13. *One combine harvester can harvest a field on its own in 12 hours. A second combine harvester can harvest the same field on its own in 18 hours. How many hours will it take both combine harvesters, working together, to harvest the whole field, if they also take a 48 minute break?*

(A) 7.2   (B) 7   (C) 9   (D) 8   (E) 10

Problem 2.14. *The width and length of the frame of a painting are in the proportion 4:5. The respective dimensions of the painting are in the proportion 3:4 and the painting is the same distance from the frame on each side. What is the ratio of the area of the frame to the area of the painting?*

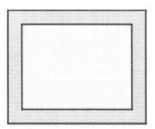

(A) $\dfrac{1}{2}$ (B) $\dfrac{2}{3}$ (C) $\dfrac{1}{3}$ (D) $\dfrac{1}{4}$ (E) $\dfrac{1}{10}$

Problem 2.15. *A boat covers the distance between piers A and B traveling downstream in two less hours than it covers the same distance traveling upstream. A raft takes 6 more hours to cover the distance from pier A to B than the boat takes to cover the same distance traveling downstream. How many hours does it take the raft to cover the distance between piers A and B?*

(A) 8   (B) 9   (C) 9.5   (D) 10   (E) 10.5

Problem 2.16. *Let AB be the diameter of a semicircle with center O. Given semicircles with diameters OA, OB and circle S, such that S is tangent to these three semicircles (see the figure). Let r be the radius of circle S. What is the value of $\dfrac{AB}{r}$?*

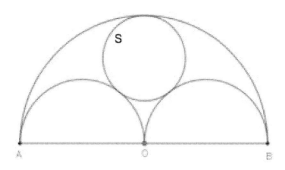

(A) 5   (B) 4   (C) 6   (D) 4.5   (E) 3

Problem 2.17. *Given a triangle with area of 10 sq. units and an inradius of 2 units. What is the area of the figure that consists of all points on the plane that are not more than 1 unit away from the sides of this triangle?*

(A) $20 + \pi$ (B) 20   (C) $20 + 2\pi$ (D) $10 + \pi$ (E) $17.5 + \pi$

Problem 2.18. *Let I be the incenter of right triangle $\triangle ABC$ with hypotenuse AB. What is the ratio of the circumradius of $\triangle ABC$ to the circumradius of $\triangle ABI$?*

(A) 2   (B) $\dfrac{2}{3}$ (C) $\dfrac{1}{2}$ (D) $\dfrac{\sqrt{2}}{2}$ (E) $\sqrt{2}$

Problem 2.19. Let A be a point on circle $\triangle$ with center C and radius 5. Circle $\triangle$ is rotated $30°$ counterclockwise around point A. What is the length of the arc of the second circle which is inside of the first circle?

(A) $\dfrac{35\pi}{6}$ (B) $\dfrac{25\pi}{6}$ (C) 4π (D) 3π (E) 5π

Problem 2.20. Let the diagonals of convex quadrilateral $ABCD$ intersect at point O. Given that the areas of $\triangle ABO$, $\triangle CBO$, $\triangle COD$ are equal to 6, 9, 18, respectively. What is the area of quadrilateral $ABCD$?

(A) 45 (B) 50 (C) 36 (D) 30 (E) 24

Problem 2.21. Let $ABCDA_1B_1C_1D_1$ be a cube with a side length of 2. Let M and N be points on sides B_1C_1 and C_1D_1, such that $BMND$ is a tangential quadrilateral. What is the length of line segment BM?

(A) $6\sqrt{2} - 4\sqrt{3}$ (B) $\sqrt{5}$ (C) $4\sqrt{2} - 2\sqrt{3}$ (D) $\dfrac{\sqrt{17}}{2}$ (E) $\sqrt{2} + \sqrt{3}$

Problem 2.22. Given 36 different rectangles of size $m \times n$, where $m \leq n \leq 8$ and m, n are positive integers. What is the probability that if two rectangles are randomly chosen, then neither of them can be covered with the other one such that the corresponding sides of the rectangles are parallel?

(A) $\dfrac{6}{35}$ (B) $\dfrac{1}{63}$ (C) $\dfrac{11}{630}$ (D) $\dfrac{1}{2}$ (E) $\dfrac{1}{5}$

Problem 2.23. Let S be the set of all positive integers from 1 to 20. How many subsets of S exist, such that each of them contains at least one prime number?

(A) 4096 (B) 2^{19} (C) 4 (D) 1044480 (E) 1024

Problem 2.24. What is the units digit of $2008^{2007^{2008}} + 2007^{2008^{2007}}$?

(A) 1 (B) 3 (C) 5 (D) 7 (E) 9

Problem 2.25. Let a square be constructed externally on each side of a regular hexagon with a side length of 1. What is the radius of the circle which passes through all the vertices of these squares that are not the vertices of the hexagon?

(A) $\sqrt{3}$ (B) 2 (C) $\dfrac{\sqrt{6} + \sqrt{2}}{2}$ (D) $\sqrt{5} + 1$ (E) $\sqrt{2}$

2.2 AMC 10 type practice test 2

Problem 2.26. *A single ice cream costs $5. At most, how many ice creams can David buy with $63?*

(A) 13   (B) 11   (C) 12   (D) 10   (E) 9

Problem 2.27. *There is $500 in each of three envelopes. The first envelope contains only $10 bills, the second one - only $20 bills, and the third one-only $50 bills. One, two, and three bills are taken out of these envelopes (in any order). What is the smallest possible amount, in dollars, that can be taken out?*

(A) 130   (B) 120   (C) 110   (D) 230   (E) 100

Problem 2.28. *What is x, if* $1 - \dfrac{1}{1 - \dfrac{1}{1+x}} = 2$?

(A) -1   (B) 0   (C) $\dfrac{1}{3}$ (D) -0.5   (E) 0.5

Problem 2.29. *Ann's father drives her from home to school every day. The school is 20 miles away from home, and Ann noticed that it took them 20, 24, and 30 minutes in the last three days to reach school. What was their average speed during last three days (in miles per hour) rounded to the nearest whole number?*

(A) 49   (B) 50   (C) 60   (D) 55   (E) 45

Problem 2.30. *What is the value of the sum of the digits of* $200920092009 \cdot 999999999999$?

(A) 90   (B) 99   (C) 81   (D) 117   (E) 108

Problem 2.31. *A circle is inscribed into a circular sector with a central angle of* $90°$. *What is the ratio of the area of the circle to the area of the sector?*

(A) $\dfrac{4}{5}$ (B) $4(3 - 2\sqrt{2})$ (C) $\dfrac{2}{3}$ (D) $\dfrac{1}{3}$ (E) $4(\sqrt{2} - 1)$

Problem 2.32. *The weight of a fish's head is 20% of its total weight and 8 kilograms less than its total weight. How much (in kilograms) does the fish weigh?*

(A) 18   (B) 9   (C) 12   (D) 16   (E) 10

Problem 2.33. *Eighty percent of the tourists in a group took a bus and the rest took a taxi. The bus fare per person is 40% less than the taxi fare per person. All together, they paid $68 for the tour transportation, and the bus fare per person is $3. How many tourists are in the group?*

(A) 15   (B) 16   (C) 18   (D) 20   (E) 10

Problem 2.34. *Positive integers* a, b, c *and 27, where* $a < b < c$, *form a geometric sequence. What is the largest possible value of* a?

(A) 1   (B) 2   (C) 8   (D) 12   (E) 16

Problem 2.35. *In right triangle* $\triangle ABC, \angle C = 90°$ *and point H is the foot of the altitude drawn from vertex C. Given* $AH = 3.6$ *and* $BH = 6.4$, *what is the perimeter of* $\triangle ABC$?

(A) 24   (B) 12   (C) 18   (D) 36   (E) 16

Problem 2.36. *If the width of a given rectangle is increased by 2 units and the length is decreased by 2 units, then the area would be 28sq. units. However, if the width of the original rectangle is decreased by 2 units and the length is increased by 2 units, then the area is 24sq. units. What is the area of the original rectangle?*

(A) 28   (B) 30   (C) 24   (D) 16   (E) 18

Problem 2.37. *Let ABC be a triangle, such that $AB = 1$ and $BC = 12$. Given that the median drawn from vertex B is BM, whose length (in units) is a positive integer. What is the length of BM?*

(A) 3   (B) 4   (C) 7   (D) 5   (E) 6

Problem 2.38. *Let m and n be positive integers, such that $2^m \cdot 3^n = a$ and $2^n \cdot 3^m = b$. What is $3^{n^2 - m^2}$?*

(A) $a^n \cdot b^m$ (B) $a^m \cdot b^{-n}$ (C) $(ab)^{n-m}$ (D) $a^{-n} \cdot b^{-m}$ (E) $a^n \cdot b^{-m}$

Problem 2.39. *Five unit squares are shown in the figure. What is the length of the side of the largest square?*

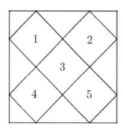

(A) 2   (B) 3   (C) $2\sqrt{2}$ (D) $\sqrt{7}$ (E) $\sqrt{6}$

Problem 2.40. *What is the value of $1^2 + 2 \cdot 2^2 + 2 \cdot 3^2 + ... + 2 \cdot 19^2 + 20^2 + 1 \cdot 2 + 2 \cdot 3 + 3 \cdot 4 + ... + 19 \cdot 20$?*

(A) 7999   (B) 8000   (C) 7200   (D) 789   (E) 800

Problem 2.41. *Points A_1, A_2, A_3, A_4, A_5 lie on the same plane. Given that the length of the line segments $A_1 A_2 = 1, A_2 A_3 = 2, A_3 A_4 = 3, A_4 A_5 = 4$. How many integer lengths are possible for line segment $A_1 A_5$?*

(A) 10   (B) 9   (C) 6   (D) 5   (E) 11

Problem 2.42. *In rectangle ABCD, $AB = 5$ and $BC = 12$. The line passing through vertex B and perpendicular to diagonal AC, intersects with side AD at point E. What is the length of line segment AE?*

(A) $1\frac{7}{12}$ (B) $2\frac{1}{12}$ (C) 2.4   (D) $2\frac{3}{5}$ (E) 2.8

Problem 2.43. *Anna, Eric, and John bought some pens. If Anna gives 50% of her pens to Eric, and after that Eric gives 50% of his pens to John, and finally John gives $\frac{100}{3}$% of his pens to Anna, then all three would have an equal number of pens. Anna had more pens than John originally by what percent?*

(A) 50   (B) 60   (C) 40   (D) 100   (E) 80

Problem 2.44. *Let a, b, c be positive integers that form a geometric sequence. Given that a has 3 divisors and c has 9 divisors. At most how many divisors can b have?*

(A) 14   (B) 9   (C) 6   (D) 8   (E) 10

Problem 2.45. A jet ski leaves port B traveling upstream towards port A, and at the same time a raft starts drifting from port A towards port B; they meet each other 3 hours after their departure. If, simultaneously, another jet ski leaves port A travelling downstream, 5mph faster than the other jet ski, the two jet skis would meet 1 hour after their departure. What is the speed (in miles per hour) of the second jet ski?

(A) 6   (B) 15   (C) 10   (D) 20   (E) 18

Problem 2.46. Three unit circles are tangent to diameter AB of the given semicircle (see the figure). Two of the unit circles are externally tangent to the third unit circle and the given semicircle. What is the length of diameter AB?

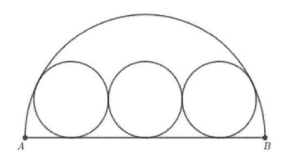

(A) $2 + 2\sqrt{5}$ (B) 6   (C) 6.8   (D) $4\sqrt{2}$ (E) $2\sqrt{3}$

Problem 2.47. Let A be the set of natural numbers from 1 to 10 inclusive, and B be the set of natural numbers from 1 to 15 inclusive. Numbers a and b are randomly chosen from sets A and B, respectively. What is the probability that ab is divisible by 6?

(A) $\dfrac{23}{50}$ (B) $\dfrac{1}{3}$ (C) $\dfrac{1}{5}$ (D) $\dfrac{11}{30}$ (E) $\dfrac{29}{150}$

Problem 2.48. The diagonals of convex quadrilateral $ABCD$ intersect at point O. The areas of $\triangle ABO$ and $\triangle CDO$ are 8 and 18, respectively. What is the smallest possible area of $ABCD$?

(A) 39   (B) 50   (C) 52   (D) 40   (E) $10\sqrt{5}$

Problem 2.49. In a regular 100-gon, a random diagonal is drawn. What is the probability that this diagonal is parallel to one of the polygon's sides?

(A) $\dfrac{1}{97}$ (B) $\dfrac{1}{2}$ (C) $\dfrac{1}{3}$ (D) $\dfrac{2}{3}$ (E) $\dfrac{48}{97}$

Problem 2.50. How many terms in the sequence $x_n = 10^n - 3^n + 2^n + 5$ are perfect squares?

(A) 0   (B) 2   (C) 1   (D) 4   (E) 3

2.3 AMC 10 type practice test 3

Problem 2.51. *There are 18 students in a 10th grade class. Five of them got 6 points on a recent test, four students got 7 points, six students got 8 points, and 3 students got 10 points. What is the average score for the test?*

(A) 7   (B) $7\frac{5}{9}$ (C) 6.5   (D) 8   (E) 9

Problem 2.52. *A square is divided into six squares (see the figure).*

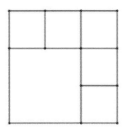

The side length of the largest of these six squares is 3. What is the side length of the original square?

(A) 9   (B) 5.5   (C) 5   (D) 4.5   (E) 6

Problem 2.53. *Arthur planned to read a book in two days, and he was supposed to read three times as many pages on the first day as on the second day. However, he read 12 fewer pages than he originally planned. In order to finish reading the book in two days, Arthur read as many pages on the second day as on the first day. How many pages does the book have?*

(A) 40   (B) 48   (C) 64   (D) 16   (E) 32

Problem 2.54. *Some trucks, each with a loading capacity of 1600 kilograms, have to transport 6.7 tonnes of sand. At least how many trucks are needed to transport the sand, if all the trucks should be loaded equally?*

(A) 4   (B) 6   (C) 5   (D) 7   (E) 8

Problem 2.55. *Let two concentric circles form a ring with an area of 36π. What is the length of the longest line segment that can be drawn in this ring (between these two circles)?*

(A) 14   (B) 10   (C) 36   (D) 12   (E) 16

Problem 2.56. *Let the operation $*$ be defined as follows: for any positive real numbers x and y, $x*y = xy - x - y$. What is x, if $x*x = 1*(1*0)$?*

(A) $\sqrt{2}+1$ (B) 1   (C) $1-\sqrt{2}$ (D) 0   (E) $\sqrt{3}-\sqrt{2}$

Problem 2.57. *In convex quadrilateral $ABCD$, $AB = 3, BC = CD = 1, \angle ABC = 150°$ and $\angle BCD = 120°$. What is the length of AD?*

(A) $\sqrt{3}$ (B) 2   (C) 1   (D) $\sqrt{3}+1$ (E) 3

Problem 2.58. *Two types of books, costing \$15 and \$26 were purchased for the school library. In all, 16 books were purchased. The total cost of the books was \$339. How many books that cost \$15 were purchased for the library?*

(A) 9   (B) 7   (C) 8   (D) 10   (E) 6

Problem 2.59. A palindrome number, such as 16061, is a number that remains the same when its digits are reversed. What is the smallest possible positive difference of two distinct four-digit palindromes?

(A) 10   (B) 11   (C) 9   (D) 110   (E) 99

Problem 2.60. In a numeric sequence $x_1 = 1, x_2 = 5$ and $x_n = |x_{n-1} - x_{n-2}|, n = 3, 4, 5, ..., 2010$. What is the remainder when x_{2010} is divided by 7?

(A) 1   (B) 2   (C) 0   (D) 4   (E) 3

Problem 2.61. The difference between positive numbers a and b is 4. The solution set to the inequality $b \leq 4 - x \leq a$ is a line segment with length 36. What is $a + b$?

(A) 4   (B) 36   (C) 2   (D) 9   (E) 18

Problem 2.62. Given three spheres, the radii of the first two spheres are 3 and 4. The sum of the total surface area of these two spheres is equal to the total surface area of the third sphere. What is the volume of the third sphere?

(A) 125π (B) 200π (C) 100π (D) $\dfrac{400\pi}{3}$ (E) $\dfrac{500\pi}{3}$

Problem 2.63. The distance between cities A and B is 320 miles. A car left City A and covered 120 miles during the first two hours of driving. After that, it stopped for half an hour and covered the rest of the trip driving at a speed of 80 mph. What was the car's average speed (in miles per hour) on the road between cities A and B?

(A) 70   (B) 64   (C) 75   (D) 72   (E) 60

Problem 2.64. Let AC be the longest side of triangle ABC. Let BH be an altitude of triangle ABC. Given that $AC = 4BH$ and $\angle C = 15°$. What is the angle measure (in degrees) of $\angle A$?

(A) 15   (B) 30   (C) 75   (D) 45   (E) 60

Problem 2.65. Each of the inhabitants of TruLi island $(A, B, C, D$ and $E)$ either always tells the truth or always lies. They once said the following:
A said: "C is a truth-teller."
B said: "D is a truth-teller."
C said: "E is a truth-teller."
D said: "B is a truth-teller."
E said: "B is a liar."
What is the value of the product of the number of truth-tellers and the number of liars on TruLi island?

(A) 6   (B) 4   (C) 0   (D) 5   (E) 3

Problem 2.66. Let D be a point on side AC of triangle ABC. Given that $AB = 6, AC = 9, CD = 5$ and $BD = 8$. What is the value of the length of side BC?

(A) $5\dfrac{1}{3}$ (B) 10   (C) 8   (D) 12   (E) 14

Problem 2.67. Ben randomly chooses a number from each of the sets: $\{1, 3, 5, 7, 9, 11\}$ and $\{2, 4, 6, 8, 10\}$. What is the probability that the sum of the chosen numbers is a multiple of 3?

(A) $\dfrac{4}{15}$ (B) $\dfrac{1}{3}$ (C) $\dfrac{1}{5}$ (D) $\dfrac{1}{4}$ (E) $\dfrac{1}{6}$

Problem 2.68. Let the edge length of a cube be 4 inches. A right circular cylinder and a square prism are removed from this cube (see the figure). Given that the radius of the base of the cylinder is 1 inch and the centers of its bases coincide with the centers of two opposite faces of the cube. The centers of the bases of the square prism coincide with the centers of two other opposite faces of the cube. The base edge length of the square prism is 2 inches and all its lateral faces are parallel to the corresponding faces of the cube. What is the value of the volume (in cubic inches) of the remaining solid?

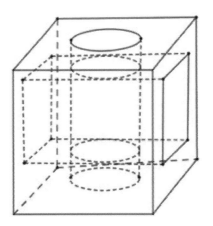

(A) $56 - 4\pi$ (B) $64 - 4\pi$ (C) 48   (D) 32   (E) $48 - 2\pi$

Problem 2.69. Let $ABCDEF$ be a convex hexagon, such that $\angle B = \angle D = \angle F = 120°$, $\angle C = 2 \cdot \angle ACE$ and $\angle A = 2 \cdot \angle CAE$. Given that the area of triangle $ACE = 10\sqrt{3}$ and the area of hexagon $ABCDEF = 11\sqrt{3}$. What is the value of $|DE - EF|$?

(A) $\sqrt{3}$ (B) 3   (C) $3\sqrt{3}$ (D) 6   (E) $6\sqrt{3}$

Problem 2.70. A fly is sitting on one of the vertices of a unit cube. What is the shortest possible length of the path it can fly, if it needs to visit each vertex of the cube?

(A) 8   (B) 6   (C) 7   (D) $6 + \sqrt{3}$ (E) $5 + \sqrt{3}$

Problem 2.71. Given nonzero numbers a, b, and c, the polynomials $x^3 + ax^2 + bx + c$, $x^2 + bx + c$, and $x^2 + ax + c$ have a common root. What is the value of $\dfrac{20a + 10b}{a+b}$?

(A) 15   (B) 10   (C) 20   (D) 14   (E) 12

Problem 2.72. Given a regular octagon, how many isosceles triangles whose vertices coincide with the vertices of the octagon exist?

(A) 16   (B) 48   (C) 64   (D) 32   (E) 24

Problem 2.73. Sam is randomly choosing n unit squares from an $n \times n$ square. The probability that the n chosen unit squares are in different columns and in different rows is $\dfrac{5}{324}$. What is n?

(A) 39   (B) 50   (C) 52   (D) 40   (E) $10\sqrt{5}$

Problem 2.74. What is the value of the sum of the last two digits of the sum $1^3 + 2^3 + ... + 2010^3$?

(A) 10   (B) 7   (C) 9   (D) 5   (E) 17

Problem 2.75. What is the smallest possible positive integer value of n, such that $0 < \{\sqrt[3]{n}\} < \dfrac{1}{99}$? Here, $\{x\}$ denotes the fractional part of a real number x.

(A) 196   (B) 210   (C) 217   (D) 187   (E) 143

2.4 AMC 10 type practice test 4

Problem 2.76. *A pen costs $0.5. When buying 10 pens you receive a discount of $1. How much will 10 pens cost?*

(A) $5 (B) $4 (C) $4.5 (D) $3 (E) $3.5

Problem 2.77. *The cost of renting a club at a driving range is $1.5. If the owner charges $0.25 for each used ball, what is the greatest number of shots that one can take for $10?*

(A) 34 (B) 35 (C) 40 (D) 36 (E) 32

Problem 2.78. *ABCD is a square with a side length of 6. Point E lies on the line segment $\overline{BC}$. What is the area of $\triangle AED$?*

(A) 30 (B) 24 (C) 12 (D) 18 (E) 6

Problem 2.79. *Among ten children on a hiking tour, any two of them have different quantities of candy. They split up equally into two groups, and it turns out that the total amount of candy in the first group is five times smaller than the total amount of candy in the second group. What is the smallest possible total amount of candy the children can have?*

(A) 70 (B) 50 (C) 60 (D) 40 (E) 30

Problem 2.80. *There are 112 apples in one box, 97 apples in a second box, and 88 apples in a third box. First, a apples from the first box were transferred to the second box, and then b apples were transferred from the second box to the third box. After that, all the boxes had an equal number of apples. What is a + b?*

(A) 11 (B) 13 (C) 20 (D) 25 (E) 24

Problem 2.81. *Aram's four children are now 1, 5, 7, and 9 years old. In how many years from now will the sum of the ages of two of his children be twice the sum of the ages of the other two children?*

(A) 1 (B) 2 (C) 3 (D) 5 (E) 10

Problem 2.82. *There are six seats in a boat. In how many ways can three girls and five boys be seated in the boat so that there are at least two girls?*

(A) 15 (B) 25 (C) 10 (D) 20 (E) 18

Problem 2.83. *What is x, if $2^{2013} : 2^x = 2^{25} \cdot 2^{1975}$?*

(A) 0 (B) $\frac{2013}{2010}$ (C) 4013 (D) 13 (E) -13

Problem 2.84. *There are 22 students in a class. 20% of the boys and 25% of the girls in the class are "A" grade students. How many "A" grade students are there in the class?*

(A) 4 (B) 2 (C) 5 (D) 3 (E) 6

Problem 2.85. *Mary read a book in four days. The number of pages she read on the second day was two times less than the number of pages she read on the first day, while the number of pages she read on the fourth day was 50% of the number of pages she read on the second day. The ratio of the number of pages she read on the third day to the number of pages she read on the fourth day is 3 : 4. What percent of the book did Mary read on the first day?*

(A) $\frac{20}{3}$ (B) 40 (C) 70 (D) $\frac{1600}{31}$ (E) 30.3

Problem 2.86. *Monday's schedule of a 10th grade class should consist of several distinct subjects. The total number of possibilities to distribute any two of the subjects for the first two classes on Monday is 20. In how many different ways can Monday's schedule be made using all the subjects?*

(A) 50   (B) 720   (C) 24   (D) 60   (E) 120

Problem 2.87. *A straight line, containing incenter I of $\triangle ABC$ and parallel to side AC, intersects sides AB and BC at points M and N, respectively. What is the perimeter of trapezoid AMNC, if MN=10 and AC=15?*

(A) 35   (B) 40   (C) 30   (D) 50   (E) 55

Problem 2.88. *How many three-digit multiples of 11 exist such that the sum of the digits is less than 11?*

(A) 16   (B) 15   (C) 10   (D) 14   (E) 12

Problem 2.89. *The sum of the number of sides, faces, and vertices of a prism is between 2012 and 2024. How many faces does the prism have?*

(A) 338   (B) 2016   (C) 337   (D) 339   (E) 350

Problem 2.90. *The lengths of the altitudes of a triangle are three consecutive terms of a geometric sequence. What is the length of the third side of the triangle, if it is given that the other two sides have lengths of 4 and 9 units?*

(A) 7   (B) 6.5   (C) 6   (D) 7.5   (E) 8

Problem 2.91. *Points $A(1,1)$ and $C(7,4)$ are reflected over the line $y=2$. What is the area of the quadrilateral whose vertices are the two given points and the two obtained points?*

(A) 10.5   (B) 13   (C) 20   (D) 17   (E) 18

Problem 2.92. *Given the following three arithmetic sequences : 1, 3, ... and 1, 4, ... and 1, 5, ... A number is randomly selected from 1 to 1000, inclusive. What is the probability that the selected number is not a term of any of the given sequences?*

(A) 0.33   (B) 0.3   (C) $\dfrac{101}{1000}$ (D) 0.5   (E) 0.333

Problem 2.93. *Convex quadrilateral ABCD, with an area of 18 and vertex coordinates $A(1,1), B(2,7), C(m,n), D(6,3)$ is drawn on the $xy-$coordinate plane. What is $m+n$?*

(A) 12   (B) 7   (C) 8   (D) 11   (E) 10

Problem 2.94. *How many natural numbers n have the following property: 2013 leaves a remainder of 13 when divided by n^2?*

(A) 6   (B) 4   (C) 10   (D) 8   (E) 5

Problem 2.95. *A square with a side length of 2 is rotated around one of its vertices by $30°$. What is the area of the part of the plane formed by the square and its rotation?*

(A) $4+\dfrac{2\pi}{3}$ (B) 4   (C) 6   (D) 2π (E) $\pi+3$

Problem 2.96. A painter painted $\frac{1}{3}$ of a wall on the first day of a job. On the second day, he painted $\frac{1}{5}$ of the remaining part of the wall, and so on. On the nth day he painted $\frac{1}{2n+1}$ of the part of the wall that remained after the $(n-1)^{th}$ day. What part of the wall still needed to be painted after the 6th day?

(A) $\frac{1}{3}$ (B) $\frac{1}{125}$ (C) $\frac{512}{3003}$ (D) $\frac{1024}{3003}$ (E) $\frac{2048}{3003}$

Problem 2.97. The centers of eight congruent spheres with radii 1 are vertices of a cube with a side length of 3. What is the radius of a sphere that is inside the cube and tangent to the eight given spheres?

(A) $1.5\sqrt{3}$ (B) 1 (C) $1.5\sqrt{3} - 1$ (D) $1.5\sqrt{3} + 1$ (E) 3

Problem 2.98. A circle passing through vertices A and C of triangle ABC, also intersects sides AB and BC at points M and N, respectively. The side lengths of triangle BMN are integers and $AB = 8, BC = 6$. What is the largest possible length of line segment MN?

(A) 10 (B) 9 (C) 8 (D) 7 (E) 6

Problem 2.99. The digits of a five-digit number are selected randomly, and the second digit from the right is even. What is the probability that the resulting five-digit number is divisible by 12?

(A) 0.5 (B) 0.1 (C) 0.4 (D) $\frac{2}{3}$ (E) $\frac{1}{3}$

Problem 2.100. At least in how many points, different from the vertices of the hexagon, the diagonals of the convex hexagon can intersect?

(A) 13 (B) 14 (C) 12 (D) 15 (E) 11

2.5 AMC 10 type practice test 5

Problem 2.101. *What is the value of* $(3^{-1} + 6^{-1} - 4^{-1})^{-1} : 5^0$?

(A) $\dfrac{1}{4}$ (B) 4 (C) 5 (D) 1 (E) $\dfrac{4}{5}$

Problem 2.102. *There are triangles and pentagons drawn on the plane, the total number of which is 11. The total sum of all the internal angles of these 11 figures is* $3420°$. *How many triangles are drawn on the plane?*

(A) 11 (B) 0 (C) 5 (D) 7 (E) 4

Problem 2.103. *Fig. 1 is made of unit length sticks. Fig. 2 is made from Fig. 1 by removing 8 equilateral triangles, with side lengths of one unit.*

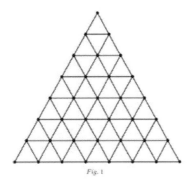
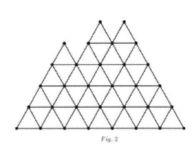

How many sticks are there in Fig. 2?

(A) 72 (B) 82 (C) 123 (D) 102 (E) 51

Problem 2.104. *A father noticed that the sum of the ages of his three children is greater by 8 than the sum of their ages 3 years ago. What is the age of his youngest child, if the children's ages are whole numbers?*

(A) 1 (B) 3 (C) 4 (D) 5 (E) 2

Problem 2.105. *There are 7 boys and 8 girls in a class. The teacher gave each of these 15 students a test with 10 problems. Each problem is scored 1 or 0 points. It is known that the average score of the girls was 7, while the average score of the class was 5.6. What was the average score of the boys?*

(A) 4.2 (B) 5 (C) 4 (D) 6 (E) 8

Problem 2.106. *The ratio of the sum of the cubes of two positive numbers to the difference of the cubes of these numbers is* $\dfrac{189}{61}$. *By what percent is the larger number greater than the smaller number?*

(A) 50 (B) 25 (C) 100 (D) 40 (E) 60

Problem 2.107. *How many two-digit numbers leave a remainder of 2 when divided by 5?*

(A) 18 (B) 19 (C) 17 (D) 90 (E) 20

Problem 2.108. *Five years ago Jane was three times Mia's age, while the sum of their ages now is 34. How old is Mia?*

(A) 6   (B) 10   (C) 13   (D) 11   (E) 8

Problem 2.109. *Points A and B lie in the first quadrant of the coordinate plane and belong to the graph of $y = \frac{1}{x^3}$. The abscissa of point A is 25% bigger than the abscissa of point B. By what percent is the ordinate of point A smaller than the ordinate of point B?*

(A) 75   (B) 25   (C) 20   (D) 50   (E) 48.8

Problem 2.110. *How many three-digit numbers have the following property: the non-negative difference of any two neighboring digits is not less than 8?*

(A) 8   (B) 6   (C) 7   (D) 9   (E) 10

Problem 2.111. *For positive numbers x and y, $x^2 - 2xy - 3y^2 = 0$. What is the value of $\frac{x^2 + 3y^2}{xy}$?*

(A) 1   (B) 4   (C) 2   (D) 5   (E) 1.5

Problem 2.112. *Let a, b, and c be positive real numbers such that the point (a,b) belongs to the graph of $y = x^2 + 2x + 2$ and the point (c, b + 3) belongs to the graph of $y = x^2 - 2x + 5$. What is c-a?*

(A) 1   (B) -2   (C) 2   (D) 0   (E) 3

Problem 2.113. *How many integers from 1 to 100 can be represented in the form $3m + 5n$, where m and n are whole numbers such that $m + n \leq 20$?*

(A) 94   (B) 96   (C) 97   (D) 95   (E) 50

Problem 2.114. *Circles with centers O_1 and O, and respective radii 1 and 5 units, are externally tangent to each other and intersect at point A. The points B and C are points on the circles such that $OB \parallel O_1C$ and the length of the smaller arc AB is equal to the smaller arc AC. What is the angular measure of $\angle AOB$ in degrees?*

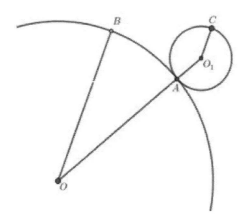

(A) 60   (B) 90   (C) 45   (D) 30   (E) 15

Problem 2.115. How many rational numbers $\frac{m}{n}$ less than 1 exist, where m and n are positive integers and $\frac{m-5}{n-5} = \frac{m^2}{n^2}$?

(A) 0   (B) 1   (C) 2   (D) 3   (E) 4

Problem 2.116. For positive numbers x and y, $y = x^3 - 2x + 2$ and $x = y^3 - 2y + 2$. What is the value of $x + y$?

(A) 1   (B) 3   (C) 4   (D) 1.5   (E) 2

Problem 2.117. What is the area of the figure consisting of all the points (x, y) in the coordinate plane, where $|x + 2| + |y + 3| \leq 5$ for each such point?

(A) 25π (B) 100   (C) 36π (D) 49π (E) 50

Problem 2.118. What is the value of the sum of all the two-digit numbers that do not have 1 as a digit?

(A) 4905   (B) 4000   (C) 4112   (D) 4312   (E) 1260

Problem 2.119. In triangle ABC, $\angle ACB = 120°$ and $AC : CB = 1 : 2$. The points D and E lie on side AB such that $\angle ACD = \angle BCE = 30°$. What is $DE : AB$?

(A) 3:10   (B) 3:5   (C) 1:2   (D) 2:3   (E) 2:5

Problem 2.120. The lengths of the sides of a rectangular prism are positive integers. The total sum of the numerical values of its volume, total surface area, and the sum of the lengths of all its sides is 2015. What is the volume of the rectangular prism?

(A) 1000   (B) 1225   (C) 1125   (D) 500   (E) 1200

Problem 2.121. Let ABCD be a tetrahedron, such that $AB = 8, AC = 4, AD = 4, BC = \sqrt{34}, BD = 4\sqrt{3}$ and $CD = 5$. What is the volume of the tetrahedron?

(A) $\frac{128}{3}$ (B) $\sqrt{39}$ (C) $\frac{128}{6}$ (D) $8\sqrt{34}$ (E) $2\sqrt{39}$

Problem 2.122. The first four digits of a twenty-digit number start with 2017, and the other digits are chosen randomly from the digits 0, 1, 2, 3, 4, 5, and 6. What is the probability that the twenty-digit number is divisible by 140?

(A) $\frac{18}{343}$ (B) $\frac{15}{343}$ (C) $\frac{12}{343}$ (D) $\frac{1}{49}$ (E) $\frac{4}{343}$

Problem 2.123. For how many values of a in the interval $(-1, 1)$ does the quadratic $x^2 + ax + 3a + 2$ have at least one integer root?

(A) 2   (B) 5   (C) 101   (D) 0   (E) 1

Problem 2.124. Let ABCD be a convex quadrilateral, such that $AB = 1, BC = 4, CD = 8$ and $AD = 7$. What is the greatest possible area of ABCD?

(A) $2\sqrt{65}$ (B) 17   (C) 18   (D) 19   (E) 18.5

Problem 2.125. Let ABC be an equilateral triangle with a side length of 2. Let M and N be randomly chosen points on line segments AB and AC, respectively. What is the probability that $MN \leq \sqrt{3}$?

(A) $\frac{\pi}{6}$ (B) $\frac{1}{2}$ (C) $\frac{1}{3}$ (D) $\frac{\pi\sqrt{3}}{6}$ (E) $\frac{6-\pi}{\pi}$

2.6 AMC 10 type practice test 6

Problem 2.126. *What is the value of the following expression*

$$\frac{1}{2} + \frac{1}{2}\cdot\frac{1}{4} + \frac{1}{2}\cdot\frac{3}{4}\cdot\frac{1}{6} + \frac{1}{2}\cdot\frac{3}{4}\cdot\frac{5}{6}\cdot\frac{1}{8} + \frac{1}{2}\cdot\frac{3}{4}\cdot\frac{5}{6}\cdot\frac{7}{8}\cdot\frac{1}{10} - 1?$$

(A) $\frac{63}{256}$ (B) $\frac{1}{256}$ (C) $-\frac{63}{256}$ (D) $-\frac{1}{256}$ (E) $\frac{65}{256}$

Problem 2.127. *Jerry solved 100 problems in n days. Each day, he solved either 5, 6 or 7 problems. What is the smallest possible value of n?*

(A) 14   (B) 15   (C) 20   (D) 16   (E) 17

Problem 2.128. *In a rectangular room with sides 5m and 6m, there are: a bookshelf, a table, a sofa, and an armchair. They respectively occupy rectangular spaces of size $0.75 \times 2.0m$, $0.8 \times 4.0m$, $0.9 \times 2.0m$, and $0.9 \times 1.0m$. What part of the room is empty?*

(A) $\frac{37}{150}$ (B) $\frac{1}{2}$ (C) $\frac{1}{3}$ (D) $\frac{113}{150}$ (E) $\frac{2}{3}$

Problem 2.129. *At 9 : 00 AM, two snails start moving simultaneously from the same point in different directions on a circular track with a circumference of 21 meters. Every hour, they change the direction in which they move. They cover a distance of $2 - (-1)^n m$ and $4 - 2(-1)^n$ after the nth hour. At what time will the snails meet each other for the first time?*

(A) 16:00   (B) 15:20   (C) 15:30   (D) 15:00   (E) 14:00

Problem 2.130. *The sum of the squares of two positive numbers is three times the product of the numbers. How many times larger is the square of the sum of these numbers than the square of their difference?*

(A) 5   (B) 3   (C) 9   (D) 2   (E) 4

Problem 2.131. *A student was assigned a test consisting of 10 problems. Each correct answer received 2 points, an incomplete answer received 1 point, and no points were given for a wrong answer. The student's total score after the test was 19. Which of the following statements is true:*

(A) The student correctly answered all 10 problems.
(B) At least one of the answers was incorrect.
(C) All the answers for all 10 problems were incomplete.
(D) Only one answer was correct.
(E) Only one answer was incomplete.

Problem 2.132. *By what percent is the circumference of a circle inscribed in a square less than the perimeter of the square?*

(A) 50%   (B) 22%   (C) 21%   (D) $\frac{100(4-\pi)}{\pi}\%$ (E) $25(4-\pi)\%$

Problem 2.133. *The participants of a mathematics conference stay in two hotels. Participants staying in the same hotel shook hands with each other, while participants staying in different hotels did not. The total number of handshakes is equal to the product of the number of participants in each hotel. What is the number of conference participants, if it is larger than 17 and less than 34?*

(A) 20   (B) 25   (C) 18   (D) 30   (E) 33

Problem 2.134. A boat starts moving downstream, and covers a distance equal to the distance that would have taken three hours if it were moving upstream. Then it moves upstream. It travels a distance moving upstream equal to the distance that would have taken two hours if it were moving downstream. What is the ratio of the boat's rate to the rate of the current, if the boat's whole trip took 5 hours?

(A) 5   (B) 4   (C) 10   (D) 2   (E) 3

Problem 2.135. Square ABCD has a side length of 12. M is an interior point of ABCD, and the distance from M to sides AB, AD, and CD is a, b, and c, respectively. How many possible points M are there, such that a, b, and c are integers and there exists a triangle with sides a, b, and c?

(A) 72   (B) 60   (C) 66   (D) 59   (E) 61

Problem 2.136. Given a cube with a side length of 1 unit. Let Φ be the solid formed by all points that are located within a distance of 1 unit from any point on the surface of the cube. What is the volume of Φ?

(A) 27   (B) $7 + 3\pi$ (C) $7 + \dfrac{4\pi}{3}$ (D) $7 + \dfrac{13\pi}{3}$ (E) $6 + 4\pi$

Problem 2.137. What is the loci of all points (x, y) for which the inequality $|x + y - 1| + |x - y + 1| + |x + y + 1| + |x - y - 1| \leq 4$ holds true?

(A) four vertices of a square   (B) a square and its inner region   (C) a triangle and its inner region   (D) three vertices of a triangle   (E) eight points

Problem 2.138. Let the number sequence x_n be the remainder when $x_{n-1}^2 + x_{n-2}^3$ is divided by 5, for n = 3, 4, ... and $x_1 = 1, x_2 = 2$. What is x_{2017}?

(A) 0   (B) 1   (C) 2   (D) 3   (E) 4

Problem 2.139. If Pablo gives 20% of his money to Mary, then Mary would have 25% more money than Pablo. If Mary gives 20% of her money to Pablo, he would have more money than Mary by what percent?

(A) 81.25%   (B) 52%   (C) $44\dfrac{24}{29}\%$ (D) $57\dfrac{11}{17}\%$ (E) 25%

Problem 2.140. Circles with radii 8 and 2 are positioned such that the distance between their centers is 3. A circle with radius 1 is randomly placed within the circle with radius 8. What is the probability that the circles with radii 1 and 2 intersect?

(A) $\dfrac{1}{16}$ (B) $\dfrac{9}{64}$ (C) $\dfrac{9}{49}$ (D) $\dfrac{1}{5}$ (E) $\dfrac{1}{2}$

Problem 2.141. A natural number is considered "nice", if at least six of its divisors are from the set of $\{1, 2, 3, 4, 5, 6, 7, 8, 9, 10\}$. What is the smallest possible value of the positive difference of two "nice" numbers?

(A) 1   (B) 3   (C) 6   (D) 4   (E) 2

Problem 2.142. All the points with integer coordinates (x, y) on the circle given by $x^2 + y^2 = 65$ form a polygon. What is the area of the polygon?

(A) 128   (B) 198   (C) 256   (D) 200   (E) 70

Problem 2.143. Mary is randomly choosing three numbers from the set of 1, 2, ..., 10. What is the probability of the event that the chosen numbers form either an arithmetic or a geometric progression?

(A) $\dfrac{23}{120}$ (B) $\dfrac{1}{6}$ (C) $\dfrac{5}{24}$ (D) $\dfrac{1}{5}$ (E) $\dfrac{1}{4}$

Problem 2.144. *Five chairs are placed around a circular table. In how many different ways can two girls and three boys be seated, such that two girls do not sit next to each other?*

(A) 120   (B) 60   (C) 30   (D) 20   (E) 24

Problem 2.145. *Let $n = \overline{a_1 a_2 \ldots a_k}$ and $T(n) = |a_1 - a_2 + \ldots + (-1)^{k-1} a_k|$. For example $T(1237) = |1 - 2 + 3 - 7| = 5$. For some natural number n, $T(n) = 4$. Which of the following values can be $T(n-1)$?*

(A) 2   (B) 9   (C) 6   (D) 1   (E) 7

Problem 2.146. *A square with a side length of x is inscribed into a triangle with side lengths of 13, 14, and 15, such that two of its vertices lie on the smallest side of the triangle. Another square, with a side length of y, is inscribed into a second congruent triangle, such that two of its vertices lie on the biggest side of the triangle. What is $\dfrac{15}{x} - \dfrac{13}{y}$?*

(A) $\dfrac{56}{195}$ (B) 1   (C) $\dfrac{1685}{1703}$ (D) $\dfrac{2}{3}$ (E) $\dfrac{3}{2}$

Problem 2.147. *Equilateral triangle ABC is inscribed in a circle with center O and radius R. A circle σ, with center O and radius $\dfrac{R\sqrt{3}}{3}$, is constructed. What is the area of the portion of triangle ABC that lies outside circle σ?*

(A) $\dfrac{R^2 \sqrt{3}}{3}$ (B) $\dfrac{3\sqrt{3} - \pi}{6} R^2$ (C) $\dfrac{3\sqrt{3} + 2\pi}{6} R^2$ (D) $\dfrac{4 - \pi}{6} R^2$ (E) $\dfrac{2\sqrt{3} - \pi}{3} R^2$

Problem 2.148. *Points with integer coordinates (i, j), where i and j are between 1 and 4 inclusive, are placed on the coordinate plane. Out of these 16 points, how many points, at most, can belong to a circle?*

(A) 3   (B) 4   (C) 5   (D) 8   (E) 10

Problem 2.149. *There exists a rational number k, such that each of the polynomials $x^3 + x^2 + kx + 2$ and $x^4 - 4x^3 + 4x^2 + (k+5)x - 3$ has an integer root larger than 1. What is k?*

(A) $\dfrac{5}{24}$ (B) $-\dfrac{3}{25}$ (C) -7   (D) 70   (E) 3

Problem 2.150. *A three-digit number is considered "fancy" if none of its digits exceed 8, and if increasing each digit by 1 results in a new three-digit number that is divisible by 11. How many possible three-digit "fancy" numbers are there?*

(A) 81   (B) 55   (C) 60   (D) 54   (E) 56

2.7 AMC 10 type practice test 7

Problem 2.151. *What is the value of $2^{3^{10}} - 8^{3^9}$?*

(A) 2   (B) 0   (C) 2020   (D) 1   (E) 1024

Problem 2.152. *How many 0 digits are there at the end of $(18! - 15!)$?*

(A) 2   (B) 3   (C) 4   (D) 5   (E) 1

Problem 2.153. *Anna and Bonita were born on the same date in different years. What is Anna's age, if two years ago she was twice as old as Bonita, and three years ago she was three times elder than Bonita.*

(A) 3   (B) 4   (C) 5   (D) 6   (E) 7

Problem 2.154. *The unit squares of a 5×12 grid are colored like a chessboard in alternating black and white colors. The diagonal of the rectangle is drawn. The intersections of the diagonal and the black squares are line segments. What is the total length, in units, of these segments?*

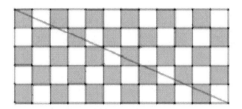

(A) 6.5   (B) 8   (C) 9   (D) 9.5   (E) 10

Problem 2.155. *Given two congruent circles on the plane with non-coincident centers. How many of the transformations below map one circle onto another?*
- *parallel translation.*
- *point symmetry.*
- *line symmetry.*
- *rotation by $30°$ angle.*

(A) 4   (B) 3   (C) 2   (D) 1   (E) 0

Problem 2.156. *What is the greatest integer that cannot be written as a sum of two composite numbers?*

(A) 101   (B) 11   (C) 111   (D) 2019   (E) 42

Problem 2.157. *How many odd positive divisors does 30^{10} have?*

(A) 121   (B) 1331   (C) 665   (D) 666   (E) 667

Problem 2.158. *How many of given quadrilaterals have an interior point that is equidistant from each line containing each side of this quadrilateral?*
- *Square.*
- *Rhombus.*
- *Parallelogram whose adjacent sides are not congruent.*
- *Isosceles trapezoid whose legs are congruent to the midsegment.*
- *Trapezoid for which the sum of the lengths of the legs is more than twice the midsegment.*

(A) 1   (B) 4   (C) 2   (D) 5   (E) 3

Problem 2.159. A geometric sequence consists of five terms. The arithmetic mean of the first four terms of this sequence is 10. The arithmetic mean of the last four terms is 30. What is the fifth term of the sequence?

(A) 40 (B) 52 (C) 64 (D) 72 (E) 81

Problem 2.160. A shop sells two-colored balls. Ten of the balls in stock are red and blue, 7 are blue and yellow, and 9 are red and yellow. Suppose n balls are chosen at random. What is the least possible value of n such that you can be certain at least 12 balls will have the same color on them?

(A) 14 (B) 16 (C) 17 (D) 15 (E) 18

Problem 2.161. Let ABC be a triangle, such that $AC = 10, BC = 17$ and $AB = 21$. Let M and N be points on side AB, such that $CM^2 = AM \cdot BM$ and $CN^2 = AN \cdot BN$ (see the figure). What is the value of the area of triangle CMN?

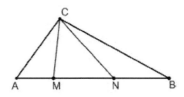

(A) 34 (B) 42 (C) 30 (D) 36 (E) $10\sqrt{3}$

Problem 2.162. How many positive integers N not exceeding 2020 have the following property: the sum of the first N positive integers divides the sum of its squares?

(A) 404 (B) 505 (C) 674 (D) 1010 (E) 1011

Problem 2.163. Let points M, N lie on sides BC, CD of parallelogram $ABCD$, respectively. Given that $\angle BAD = 85°, \angle MAN = 20°$ and $\angle MNA = 10°$. Let ME be the angle bisector in $\triangle AMN$. What is the angle measure of $\angle BEC$?

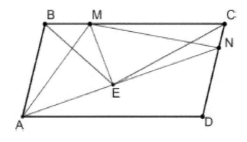

(A) 100 (B) 120 (C) 90 (D) 150 (E) 160

Problem 2.164. Consider four congruent circles on a plane, such that they are pairwise non-concentric and they divide the plane in n parts. How many values of n are possible?

(A) 10 (B) 9 (C) 8 (D) 7 (E) 6

Problem 2.165. What is the least possible value of the expression $(x+4)(x+5) + \dfrac{10^6}{x(x+9)}$, where x is a positive real number?

(A) 2017 (B) 2018 (C) 2019 (D) 2020 (E) 2021

Problem 2.166. *Let x and y be randomly chosen numbers from $[0,1]$. What is the probability that the following inequality holds true?*
$$|x - 0.5| + |y - 1.5| \leq 1.$$

(A) $\dfrac{1}{4}$ (B) $\dfrac{3}{4}$ (C) $\dfrac{1}{6}$ (D) $\dfrac{5}{6}$ (E) $\dfrac{1}{2}$

Problem 2.167. *Given 22 circles, such that 21 of them have radius 1 (see the figure). Given also that any two circles that have a common point are pairwise tangent. What is the value of the circumference of the largest circle?*

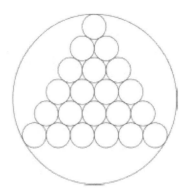

(A) $\dfrac{2\pi}{3}(10\sqrt{3}+3)$ (B) $\dfrac{20}{3}\pi$ (C) $\dfrac{20\sqrt{3}}{3}\pi$ (D) 13π (E) 20π

Problem 2.168. *How many seven-digit numbers contain 3 ones, 2 twos, 2 threes and no two neighboring digits are ones?*

(A) 90 (B) 120 (C) 60 (D) 100 (E) 150

Problem 2.169. *A rectangular 2×10 grid is randomly covered by ten 1×2 rectangles (dominos). What is the probability that the two squares in the fifth column from the left will be covered by different dominos?*

(A) $\dfrac{64}{89}$ (B) $\dfrac{49}{89}$ (C) $\dfrac{25}{89}$ (D) $\dfrac{40}{89}$ (E) $\dfrac{1}{2}$

Problem 2.170. *Let the vertices of triangle ABC lie on the surface of a sphere with a radius of 13 (see the figure). Given that $AB = 5$ and $m\angle ACB = 30°$. What is the value of the distance from the center of the sphere to the plane containing triangle ABC?*

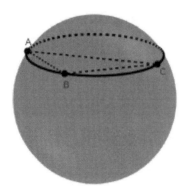

(A) 10 (B) 11 (C) 12 (D) 6 (E) 8

Problem 2.171. Let $f(x)$ be a monic polynomial function of degree five, such that

$$f(2021) = -2, f(2022) = 1, f(2023) = 2, f(2024) = 1, f(2025) = -2.$$

What is the value of $f(2020)$?

(A) -5   (B) -12   (C) -127   (D) -210   (E) -2100

Problem 2.172. Let (a_n) be a number sequence defined as follows: $a_1 = a_2 = 1, a_3 = \dfrac{1}{2}$ and $a_{n+1} = \dfrac{a_n^3 a_{n-2}}{a_{n-1}^3 + a_{n-2}a_{n-1}a_n}$, where $n = 3, 4, \ldots$ What is the value of

$$S_{10} = \frac{a_1}{a_2} + 2\frac{a_2}{a_3} + 3\frac{a_3}{a_4} + \ldots + 10\frac{a_{10}}{a_{11}}.$$

(A) 11!   (B) 11! $- 1$ (C) 10! $+ 3$ (D) 10! $+ 11$ (E) $10 \cdot 11$

Problem 2.173. Let a, b, c be integers, such that $x^2 + 3x + 2 \leq ax^2 + bx + c \leq 2x^2 + 12x + 25$ double inequality holds true for all values of x. What is the value of the sum of all possible values of c?

(A) 60   (B) 61   (C) 80   (D) 81   (E) 90

Problem 2.174. A league of soccer teams participate in a tournament. Any two teams play each other exactly once. By the end of the tournament, exactly n games end in a tie and the total points gained in the tournament is 2019. What is the value of the sum of all possible values of n? (Note: A winning team gets 3 points, a losing team gets 0 points, while a tie is 1 point.)

(A) 2019   (B) 2050   (C) 2150   (D) 4038   (E) 4080

Problem 2.175. Let n be a positive integers such that $\dfrac{(n!)^2}{(n+3)!}$ is also a positive integer. What is the smallest possible number of divisors of $(n+1)(n+2)(n+3)$?

(A) 24   (B) 30   (C) 36   (D) 12   (E) 20

2.8 AMC 10 type practice test 8

Problem 2.176. *Some of the students in the class are students in a math circle as well. Most of the students in the class participated in the AMC 10 test. It appeared that 7/8 of the math circle students and 5/6 of students not attending the math circle participated in the AMC 10. What part of the class participated in the AMC 10, if the total number of students in the class is not more than 19?*

(A) $\frac{1}{2}$ (B) $\frac{2}{3}$ (C) $\frac{6}{7}$ (D) $\frac{3}{4}$ (E) $\frac{8}{9}$

Problem 2.177. *There are 360 books in the library. 20% of the books that are about mathematics and 50% of the books that are not about mathematics are hard cover. Given that 0.3 part of the books in the library are hard cover, how many books about mathematics are in the library?*

(A) 240   (B) 120   (C) 180   (D) 270   (E) 300

Problem 2.178. *Given that the mean of $30, 31, ..., 29 + 2n$ is equal to the mean of $100, 101, ..., 99 + n$. What is the value of the sum of the digits of n?*

(A) 10   (B) 5   (C) 8   (D) 7   (E) 6

Problem 2.179. *Two boxes contain an equal number of pens. Given that all pens are either white or black. The ratio of white to black pens in the first box is $7 : 3$ and $3 : 2$ in the second box. If the total number of black pens is 35, how many white pens are in the first box?*

(A) 20   (B) 25   (C) 30   (D) 35   (E) 40

Problem 2.180. *For how many of the following quadrilaterals does there never exist a point (on the plane of that quadrilateral) that is equidistant from all four sides of that quadrilateral?*
- *A rhombus that is not a square.*
- *A parallelogram that is not a rectangle or a rhombus.*
- *An isosceles trapezoid that is not a parallelogram.*
- *Not isosceles trapezoid.*
- *A quadrilateral that is neither a parallelogram nor a trapezoid.*

(A) 1   (B) 2   (C) 3   (D) 4   (E) 5

Problem 2.181. *Given that $n!, (n+1)!$ and $n!(n+19)$ form an arithmetic sequence where n is a positive integer. What is the product of all digits of n?*

(A) 18   (B) 24   (C) 30   (D) 3   (E) 8

Problem 2.182. *Suppose that lines $ax + by = c, (a+2)x + (b-3)y = c+9$ and $(a+1)x + (b-1)y = c+2$ are concurrent at point M. What is the value of the sum of the coordinates of M?*

(A) 0   (B) 1   (C) -1   (D) -2   (E) -8

Problem 2.183. *Positive integers m and n are such that both the sum and the difference of $mn+m+n-53$ and $9n - 6m + 3$ are prime numbers. What is the value of the sum of the digits of the product mn?*

(A) 7   (B) 9   (C) 10   (D) 13   (E) 15

Problem 2.184. *Given 1×2 and 3×6 rectangles (see the figure). What is the value of the area of $\triangle ABC$?*

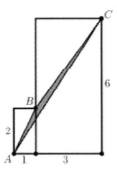

(A) 2 (B) 3 (C) 1 (D) 1.5 (E) 0.5

Problem 2.185. *Let function f is defined by $f(x) = \lfloor 3x \rfloor - \lfloor 2x \rfloor - \lfloor x \rfloor$ for all real numbers x, where $\lfloor r \rfloor$ denotes the greatest integer that is less than or equal to the real number r. What is the range of f?*

(A) $\{0\}$ (B) $\{1\}$ (C) $\{0,1\}$ (D) $\{0,1,2\}$ (E) $\{0,2\}$

Problem 2.186. *Let the lengths of two of the sides of $\triangle ABC$ be 6, 8 and its area be 24. What is the value of the perimeter of $\triangle ABC$?*

(A) 20 (B) 24 (C) 22 (D) 27 (E) 26

Problem 2.187. *For how many positive integers n, both fractions $\dfrac{20n+19}{n+817}$ and $\dfrac{19n+61}{n+817}$ are positive integers too?*

(A) 0 (B) 1 (C) 3 (D) 5 (E) 11

Problem 2.188. *Suppose you write the set $\{4000, 4001, 4002, ...5000\}$ in base 10. Then you recorded this set in base 5 and found the number in the new set whose digits have the least sum. What is the value of the sum of the digits of this number?*

(A) 7 (B) 6 (C) 5 (D) 4 (E) 3

Problem 2.189. *Given that $19! = 121,645,100,408,ab2,000$, what is the largest prime factor of the number $\overline{ab2}$?*

(A) 11 (B) 7 (C) 13 (D) 17 (E) 5

Problem 2.190. *Given that the area of the triangle with sides a, b, c and the area of the triangle with sides a, b, d, where $d \neq c$, are equal. Given also that $c^2 + d^2 = 2020$. What is the value of $a^2 + b^2$?*

(A) 505 (B) 1010 (C) 101 (D) 400 (E) 640

Problem 2.191. *Let points D and E lie on sides AC and BC of triangle ABC, respectively. Given that $\angle BAC = 30°, \angle ACB = 15°, AB = BD$ and $DE = EC$. What is the ratio $AD : CD$?*

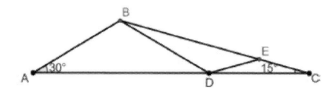

(A) $\sqrt{3} : 1$ (B) $1 : \sqrt{3}$ (C) $1 : 2$ (D) $2 : 1$ (E) $1 : 3$

Problem 2.192. *Let A denote the smallest positive integer that is divisible by 36 and whose base 10 representation consists of only 4's and 9's, with at least one of each. What is the value of the remainder of A after a division by 11?*

(A) 1 (B) 3 (C) 5 (D) 8 (E) 10

Problem 2.193. *A hotel with infinite number of rooms has room numbers labeled with the positive integers 1, 2, 3,... A guest is assigned to room i with probability $\frac{1}{2^i}$. What is the probability that the positive difference of the room numbers assigned to two random guests is 2?*

(A) $\frac{1}{3}$ (B) $\frac{2}{3}$ (C) $\frac{1}{12}$ (D) $\frac{1}{6}$ (E) $\frac{1}{4}$

Problem 2.194. *Given that point $M(x_0, y_0)$ and circle σ lie on the same plane. Secants MB and MD are drawn such that the rays BM, DM meet σ at points A, C, respectively. Given that $AB = CD, A(2,4), B(3,6)$ and $C(0,2)$. What is the value of $4x_0 + y_0$?*

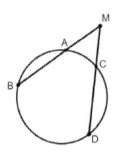

(A) 2 (B) 3 (C) 8 (D) 0 (E) 5

Problem 2.195. *Given that 10% of the numbers in the set $\{1, 2, ..., n\}$ is divisible by 9. What is the total number of all possible values of n?*

(A) 2 (B) 9 (C) 13 (D) ∞ (E) 8

Problem 2.196. *Let a be the least possible value of x, such that the values of the functions $y = \frac{5}{6}x + \frac{1}{3}$ and $y = \frac{25}{16}x - \frac{1}{2}$ are positive integers. What is the value of the sum of the digits of a^3?*

(A) 1 (B) 2 (C) 3 (D) 6 (E) 8

Problem 2.197. Let ABC be a triangle, such that $AC = 3, BC = 5, AB = 7$ and its each side is the diameter of a semicircle (see the figure). The areas of the grey figures are labeled S_1, S_2 and S_3. What is the value of $S_1 + S_3 - S_2$?

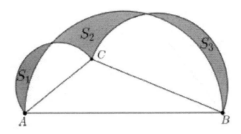

(A) $\dfrac{15}{8}(2\sqrt{3} - \pi)$ (B) $\dfrac{15}{4}(2\sqrt{3} - \pi)$ (C) $\dfrac{\pi}{16} - \dfrac{\sqrt{3}}{18}$ (D) 4   (E) $\pi + 2$

Problem 2.198. How many numbers $\overline{a_1 a_2 ... a_{10}}$ with nonzero digits exist, such that each of the numbers $\overline{a_1 a_2 a_3}, \overline{a_2 a_3 a_4}, ..., \overline{a_8 a_9 a_{10}}$ is not divisible by 3?

(A) $2^8 \cdot 3^{10}$ (B) $2^8 \cdot 3^{12}$ (C) $2^{10} \cdot 3^{10}$ (D) $2^{12} \cdot 3^{12}$ (E) $2^{12} \cdot 3^8$

Problem 2.199. A random four-digit number is chosen. What is the probability that the digits of the chosen number are four consecutive numbers?

(A) $\dfrac{1}{30}$ (B) $\dfrac{9}{500}$ (C) $\dfrac{7}{300}$ (D) $\dfrac{11}{450}$ (E) $\dfrac{7}{80}$

Problem 2.200. Let sequence (x_n) be defined as follows: $x_1 = 1$ and $x_{n+1} = x_n + \dfrac{1}{x_n}$. What is the value of the sum of the digits of the smallest possible n, such that $x_n > 8$?

(A) 5   (B) 12   (C) 16   (D) 21   (E) 24

2.9 AMC 10 type practice test 9

Problem 2.201. *What is the value of the expression $(((1^{-1} - 2^{-1})^{-2} - 3)^{-2} - 4)^{-2}$?*

(A) 9   (B) 3   (C) 1   (D) $\frac{1}{3}$ (E) $\frac{1}{9}$

Problem 2.202. *On the entrance of the shop is written $20\% + 20\%$ discount. It means, that the price is decreased by 20% and then the new price is again decreased by 20%. By what percent can the $20\% + 20\%$ discount be replaced?*

(A) 40   (B) 38   (C) 36   (D) 35   (E) 34

Problem 2.203. *For how many different values of variable x the expression $\dfrac{1}{2 - \dfrac{3}{4 - \dfrac{1}{x^2}}}$ is not defined?*

(A) 1   (B) 2   (C) 3   (D) 4   (E) 5

Problem 2.204. *Given that one of the numbers 1, 2, ..., 9 is equal to the arithmetic mean of the other eight numbers. What is the value of the sum of the digits of the sum of those eight number?*

(A) 2   (B) 3   (C) 4   (D) 6   (E) 11

Problem 2.205. *Let n be a positive integer. What is the total number of all the elements of the set $\left\{1, \frac{1}{2}, \frac{1}{3}, ..., \frac{1}{n}, ...\right\}$, such that each of them is a solution of the inequality $100x^2 - 25x + 1 < 0$?*

(A) 13   (B) 14   (C) 15   (D) 16   (E) 19

Problem 2.206. *A student took the bus to Harvard from Newton. He got into the bus at the first station and got off at the last station. The student noticed, that 10% of all passengers got into the bus at the first station and 60% of all passengers got off at the last station. The student also noticed, that in between the first and the last stations 8 people got off the bus. How many people got into the bus at the first station?*

(A) 1   (B) 3   (C) 10   (D) 5   (E) 2

Problem 2.207. *For which value of m do lines $y = x - 1$, $y = 3x - 5$ and $y = mx - 41$ intersect at one point?*

(A) 5   (B) 10   (C) 19   (D) 21   (E) 22

Problem 2.208. *One of the shops sells 450 grams of a candy for 5\$, the second shop sells 500 grams of the same candy for 6\$. By how many percent the candy sold in the second shop is more expensive than the candy sold in the first shop?*

(A) 12   (B) 10   (C) 9   (D) 8   (E) 5

Problem 2.209. *Let a and b be real numbers, such that $a + b = 1$ and $(a^2 + b)(b^2 + a) = 2019$. What is the value of the expression $(a^2 + 1)(b^2 + 1)$?*

(A) 100   (B) 200   (C) 2018   (D) 2019   (E) 2020

Problem 2.210. *Let D be a given point on side AC triangle ABC, such that $AD = 1$, $CD = 7$, $BD = 4$ and $\angle ADB = 90°$. What is the length of the median drawn from vertex B of triangle ABC?*

(A) $3\sqrt{2}$ (B) $2\sqrt{5}$ (C) 5   (D) $3\sqrt{3}$ (E) 6

Problem 2.211. Let $\overline{ab}$, $\overline{cd}$, $\overline{ac}$, $\overline{bd}$ be two-digit numbers, such that $\overline{ab} + \overline{cd} = \overline{ac} + \overline{bd}$. What is the total number of all possible quadruples (a, b, c, d)?

(A) 810   (B) 729   (C) 700   (D) 500   (E) 100

Problem 2.212. Given three pairwise different positive odd integers, such that the sum of any two of them is greater at least by 5 than the third number. What is the smallest possible value of the sum of these three numbers?

(A) 17   (B) 23   (C) 25   (D) 27   (E) 29

Problem 2.213. Let two circles of radii 6 be tangent to each other, such that each of these circles is tangent to bases AD and BC and to one of the legs of trapezoid $ABCD$. Given that $AB = 13$ and $CD = 20$. What is the area of trapezoid $ABCD$?

(A) 198   (B) 396   (C) 300   (D) 342   (E) 210

Problem 2.214. What is the value of the following expression?

$$\frac{3}{1\cdot 2} - \frac{5}{2\cdot 3} + \frac{7}{3\cdot 4} - \frac{9}{4\cdot 5} + \frac{11}{5\cdot 6} - \frac{13}{6\cdot 7} + \frac{15}{7\cdot 8} - \frac{17}{8\cdot 9} + \frac{19}{9\cdot 10}.$$

(A) 1.1   (B) 0.9   (C) 0.75   (D) 0.6   (E) 0.5

Problem 2.215. The numbers -5, -4, -3, -2, -1, 0, 1, 2, 3, 4, 5 are written on the board. James can pick any two of these numbers, for example numbers a and b, afterward James writes the value of the expression $ab - a - b + 2$ on the board. Then he erases numbers a and b from the board. For the next step, James repeats the same with the remaining 10 numbers. After the 10th step, only one number is written on the board. What is the value of the last number left on the board?

(A) -120   (B) -5   (C) 0   (D) 1   (E) 4

Problem 2.216. Let m be a positive integer and n be a non-negative integer. What is the total number of all (m, n) pairs, such that

$$\sum_{i=1}^{m!} i = 2^{2n} + 2^n + 1?$$

(A) 1   (B) 2   (C) 3   (D) 6   (E) 7

Problem 2.217. Let a and b be positive integers and $a > b$. Consider points $A(a, b)$, $B(a - b, a)$, $O(0, 0)$ on the coordinate plane. Given that the area of triangle ABO is 96. What is the value of $a + b$?

(A) 28   (B) 27   (C) 26   (D) 25   (E) 24

Problem 2.218. Let AD is the bisector of angle BAC of triangle ABC. Given that $BD = 3$, $CD = 5$ and $AC - AB = 4$. What is the value of the measure (in degrees) of angle ABC?

(A) 120   (B) 100   (C) 90   (D) 75   (E) 60

Problem 2.219. What is the total number of all triples (p, q, r) of prime numbers, such that $p < q < r$ and each of the numbers $q - p$, $r - q$, $r - p$ is not a composite number?

(A) 0   (B) 1   (C) 2   (D) 6   (E) 12

Problem 2.220. *Given a rectangle ABCD and circles σ_1, σ_2, σ_3, σ_4, each with radius $r = \sqrt{2-\sqrt{3}}$ (see the figure). Let each of the circles σ_1, σ_3 touch each of the circles σ_2, σ_4. Given that circles σ_2, σ_4 touch each other and sides AB, CD, respectively. Given also that circles σ_1, σ_3 touch respectively sides AB, AD and BC, CD. What is the area of rectangle ABCD?*

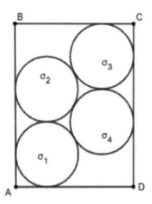

(A) $2\sqrt{3}(2-\sqrt{3})$ (B) 4   (C) $2\sqrt{3}$ (D) 5   (E) $3+2\sqrt{3}$

Problem 2.221. *What is the total number of all seven-digit numbers consisting only of the digits 1, 2, 3, 4, 5, 6, 7 (in any order), such that the digits are not repetitive and each is divisible by 11?*

(A) 120   (B) 240   (C) 320   (D) 400   (E) 576

Problem 2.222. *Let $\overline{abc}$ be the smallest three-digit number, such that when it is divided by the product of its digits the remainder is equal to the sum of its digits. What is the quotient of that division?*

(A) 11   (B) 12   (C) 13   (D) 14   (E) 15

Problem 2.223. *Points M, N, P, Q are given correspondingly on sides AB, BC, CD, AD of the rectangle ABCD. Let R be the intersection point of line segments MP and QN. Given that $\angle MRN = 90°$, $MR = 21$, $NR = 27$, $QR = 33$ and $\dfrac{AB}{BC} = \dfrac{3}{4}$. What is the length of PR?*

(A) 39   (B) $42\dfrac{3}{7}$ (C) 45   (D) 59   (E) 61

Problem 2.224. *There were 8 chairs at a round table and on each chair was sitting one person. After the break, each person randomly chooses to sit either on the chair that he/she was sitting before the break or on one of its neighboring two chairs (again on each chair is sitting one person). What is the probability of the event that after the break exactly two persons are sitting on their previous chairs?*

(A) $\dfrac{1}{3}$ (B) $\dfrac{8}{39}$ (C) $\dfrac{1}{2}$ (D) $\dfrac{16}{53}$ (E) $\dfrac{16}{49}$

Problem 2.225. *What is the value of the sum of all positive integers n, such that for each such n the value of the expression $\dfrac{n(n+7)}{20}$ is equal to some prime number raised to a positive integer power?*

(A) 18   (B) 33   (C) 38   (D) 61   (E) 66

2.10 AMC 10 type practice test 10

Problem 2.226. *What is the value of the following expression?*

$$\sqrt[6]{2^5 \cdot \sqrt[3]{2^6 \cdot \sqrt[8]{2^7 \cdot \sqrt[9]{2^8 \cdot \sqrt[10]{2^{10}}}}}}.$$

(A) $\sqrt[6]{2}$ (B) 2 (C) 4 (D) 8 (E) 1024

Problem 2.227. *The arithmetic mean of $a, 3, 4$ is 5 more than the arithmetic mean of $b, -1, 4$. What is the value of non-negative difference of a and b?*

(A) 18 (B) 5 (C) 11 (D) 0 (E) 2

Problem 2.228. *What is the value of x, such that the following equation holds true?*

$$x - \frac{1}{6} = \frac{1}{1 \cdot 2} + \frac{1}{2 \cdot 3} + \frac{1}{3 \cdot 4} + \frac{1}{4 \cdot 5} + \frac{1}{5 \cdot 6}.$$

(A) 0 (B) $\frac{1}{6}$ (C) $\frac{31}{30}$ (D) $\frac{1}{2}$ (E) 1

Problem 2.229. *What is the value of the smallest positive integer n, such that the value of the expression $\frac{n-2}{5} - \frac{n+3}{6}$ is also a positive integer?*

(A) 1 (B) 10 (C) 27 (D) 57 (E) 87

Problem 2.230. *John got 150000\$ 30-year fixed 4.8 % interest rate mortgage loan from the bank. Each year in total he should pay back the same amount of money (principal+interest) and each year 4.8% interest is calculated on the principal left after the payment of the previous year. In total how much money should John pay to the bank in the end of the second year?*

(A) 11625 (B) 11960 (C) 12000 (D) 15000 (E) 54800

Problem 2.231. *What is the value of the following expression?*

$$\left(\frac{1}{7} + \frac{1}{21} - \frac{1}{14} - \frac{1}{28} - \frac{1}{12}\right)^2 + \left(\frac{1}{5} + \frac{1}{10} + \frac{1}{20} - \frac{1}{4} - \frac{1}{15} - \frac{1}{30}\right)^2 + \left(\frac{1}{8} + \frac{1}{24} - \frac{1}{9} - \frac{1}{18}\right)^2 + \left(\frac{1}{2} + \frac{1}{3} + \frac{1}{6} - 1\right)^2.$$

(A) $-\frac{1}{2}$ (B) $-\frac{1}{3}$ (C) 0 (D) $\frac{1}{3}$ (E) 1

Problem 2.232. *What is the value of the sum of all distinct solutions of the equation $((x-4)^2 - 3)^2 = 9$?*

(A) 16 (B) 15 (C) 14 (D) 12 (E) 0

Problem 2.233. *Two numbers are randomly chosen from the set of all two-digit numbers. What is the probability that the positive difference of two chosen numbers is also a two-digit number?*

(A) $\frac{36}{89}$ (B) $\frac{2}{5}$ (C) $\frac{72}{89}$ (D) $\frac{4}{5}$ (E) $\frac{9}{10}$

Problem 2.234. *A rectangular prism is called "beautiful" if its three measurements are positive integers. The rectangular prism M is divided into eight "beautiful" rectangular prisms with three planes parallel to its faces. Given that out of these eight "beautiful" rectangular prisms the volumes of four of them are equal to 1, 2, 3, and 5 What is the surface area of the rectangular prism M?*

(A) 108 (B) 84 (C) 72 (D) 60 (E) 30

Problem 2.235. Let number a be the smallest value of variable x, such that the values of functions $y = \frac{2}{3}x - \frac{1}{3}$ and $y = \frac{3}{4}x - \frac{1}{4}$ are positive integers. What is the value of the sum of all digits of a?

(A) 2   (B) 3   (C) 4   (D) 1   (E) 5

Problem 2.236. Steven planned to solve some number of problems in 3 days. On the first day, he solved $\frac{1}{3}$ of all problems. On the second and the third days, he solved respectively $\frac{3}{4}$ and $\frac{5}{6}$ of problems. It turned out that the number of problems he solved was 6 times more than the number of problems he did not solved. What percent of all problems did Steven solve on the second day?

(A) 25   (B) $\frac{200}{7}$ (C) $\frac{300}{7}$ (D) 50   (E) 51

Problem 2.237. Given a triangle ABC. Let CD be the angle bisector of $\angle ACB$. Given that $AC = CD$, $\angle ACB = 108°$ and $\angle A = n \cdot \angle B$. What is the value of n?

(A) 3   (B) 5   (C) 6   (D) 7   (E) 10

Problem 2.238. A positive divisor of $10!$ is randomly chosen. What is the probability that the chosen divisor is not divisible by 3?

(A) $\frac{2}{3}$ (B) $\frac{1}{3}$ (C) $\frac{1}{4}$ (D) $\frac{1}{5}$ (E) $\frac{1}{6}$

Problem 2.239. Let $P(x) = (x-2)(x-4)(x-6)(x-8)$. What is the total number of all possible pairs of integers (m, n), such that $P(m) + P(n) < 0$.

(A) 36   (B) 35   (C) 34   (D) 33   (E) 32

Problem 2.240. What is the value of the following expression?

$$\frac{1^2 + 5 \cdot 1 + 4}{1^2 + 5 \cdot 1 + 6} \cdot \frac{2^2 + 5 \cdot 2 + 4}{2^2 + 5 \cdot 2 + 6} \cdot \ldots \cdot \frac{98^2 + 5 \cdot 98 + 4}{98^2 + 5 \cdot 98 + 6}.$$

(A) 0.35   (B) 0.4   (C) 0.5   (D) 0.51   (E) 0.61

Problem 2.241. Natural numbers from 1 to 2020 are written in a row: $123456789101112\ldots20192020$. How many times is number 12 repeated in this multi-digit number?

(A) 155   (B) 105   (C) 165   (D) 150   (E) 175

Problem 2.242. Given a point M inside of the right triangle with legs AC and BC. Given that $\angle MAC = \angle MCA = 30°$ and $\frac{AC}{BC} = \frac{\sqrt{3}}{2}$. What is the value of the measure (in degrees) of $\angle AMB$?

(A) 120   (B) 135   (C) 150   (D) 160   (E) 170

Problem 2.243. Let u and v be positive numbers, such that $|u - v| \geq 1$. What is the smallest possible value of the expression $uv + \frac{u}{v} + \frac{v}{u}$?

(A) 3   (B) 3.5   (C) 4   (D) $\sqrt{35}$ (E) 6

Problem 2.244. What is the total number of all three-digit numbers, not containing any zero digit, for which there is a digit such that after erasing that digit the obtained two-digit number is divisible by 3? For example three-digit numbers 121 and 123 satisfy these conditions.

(A) 120   (B) 159   (C) 729   (D) 540   (E) 513

Problem 2.245. Let $ABCD$ be a rhombus, such that $\angle A = 45°$ and $AC = 13$. Assume that ray BD intersects the circumcircle of triangle ABC at point E. What is the value of the length of line segment DE?

(A) 12   (B) 13   (C) $13\sqrt{2}$ (D) 20   (E) 26

Problem 2.246. Two-digit number is called "beautiful", if it does not end with 0 and is divisible by the sum of its digits. What is the value of the sum of all "beautiful" numbers?

(A) 627   (B) 507   (C) 417   (D) 330   (E) 210

Problem 2.247. An ant moves from the bottom left corner to the top right corner of 4×8 rectangular grid. It can move only on the sides of unit cells, such that one move is either going up 1 unit or going down 1 unit or going right 1 unit. Given that the ant cannot pass twice the same side of any unit cell. What is the probability of the event that the ant passes through the center of symmetry of given 4×8 rectangular grid?

(A) $\dfrac{1}{2}$ (B) $\dfrac{8}{25}$ (C) $\dfrac{2}{3}$ (D) $\dfrac{17}{25}$ (E) $\dfrac{16}{25}$

Problem 2.248. 8 soccer teams participated in a tournament. Each two teams played with each other only once (a winning team gets 3 points, a losing team gets 0 points, while a draw is 1 point for each team). Given that in the end the difference of the sum of the points of the teams at the first four places and the sum of the points of the teams at the last four places is equal to 54. How many points does the team at the last place has?

(A) 0   (B) 1   (C) 2   (D) 3   (E) 10

Problem 2.249. 8 rectangular prisms with sizes $1 \times 1 \times 2$ must be placed in a rectangular prism box with size $2 \times 2 \times 4$. Given that all faces of the rectangular prism box are colored in different colors. In how many ways is it possible to do that?

(A) 28   (B) 40   (C) 64   (D) 80   (E) 100

Problem 2.250. At least how many numbers should be erased from the list of numbers $1, \dfrac{1}{2}, \dfrac{1}{3}, ..., \dfrac{1}{30}$ in order to be able to split the rest of the numbers into two groups, such that the sum of all elements of each group are equal to each other?

(A) 8   (B) 11   (C) 12   (D) 14   (E) 16

2.11 AMC 10 type practice test 11

Problem 2.251. What is the value of the expression $\dfrac{2020! + 2018!}{2017! \cdot (2019^3 - 1)}$?

(A) 2020 (B) 1 (C) $\dfrac{1}{2017}$ (D) $\dfrac{1}{2018}$ (E) $\dfrac{1}{2020}$

Problem 2.252. Which of the following statements is true? The sum of four consecutive integers can be:

(A) 0 (B) Odd number (C) Multiple of 4 (D) A square of an integer (E) Even number

Problem 2.253. The median of numbers $1, 2, x, 13$ is equal to 3. What is the value of the mean of these numbers?

(A) 5 (B) 6 (C) 10 (D) 11 (E) 12

Problem 2.254. Let a and b be integers, such that $a \cdot b = 2020$. What is the total number of all such pairs (a, b)?

(A) 10 (B) 11 (C) 12 (D) 24 (E) 30

Problem 2.255. 23 students sit in three rows in the classroom (there is at least one student in each row). Given that 20%, 25% and 10% of the students sitting respectively in the first, second and third row play basketball. In total, how many students play basketball?

(A) 3 (B) 4 (C) 5 (D) 6 (E) 7

Problem 2.256. Let $ABCD$ be a square and $AMNK$ be a rectangle, such that their perimeters are equal (see the figure). Which of the following statements holds true?

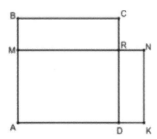

(A) $RC < RN$ (B) $RC > RN$ (C) $Area(MBCR) < Area(DRNK)$ (D) $Area(ABCD) = Area(AMNK)$ (E) $Area(ABCD) > Area(AMNK)$

Problem 2.257. Four numbers are chosen out of five consecutive integers. The sum of chosen numbers is equal to 2020. What is the value of the sum of these five consecutive integers?

(A) 1000 (B) 2000 (C) 2500 (D) 2525 (E) 3225

Problem 2.258. A paper rectangle is cut into three rectangles. The sum of the perimeters of those three rectangles is 32 and the perimeter of the initial paper rectangle is 16. What is the value of the area of the initial paper rectangle?

(A) 7 (B) 12 (C) 15 (D) 15.5 (E) 16

Problem 2.259. *Three sisters together bought three bracelets of different colors. In how many different ways can they wear all these three bracelets?*

(A) 27 (B) 48 (C) 216 (D) 264 (E) 336

Problem 2.260. *Let BE be a median of triangle ABC and CF be a median of triangle BEC. Given that $AF = AE$ and $CF = 20$. What is the length of line segment AB?*

(A) 16 (B) 17 (C) 18 (D) 19 (E) 20

Problem 2.261. *The factory did not work on every Saturday and Sunday of February. On the nth day of each week it produced $(6-n)^2$ devices, where $n \in \{1,2,3,4,5\}$. Given that in total the factory produced 236 devices during the month of February. Which day of the week was the last day of that February?*

(A) Monday (B) Tuesday (C) Wednesday (D) Thursday (E) Friday

Problem 2.262. *The probability that it will not rain for two consecutive days is $\frac{1}{6}$, the probability that it will rain on both days is $\frac{1}{3}$. The probability that it will rain only on one of those two days is $\frac{1}{2}$. What is the probability that it will rain exactly two days out of six consecutive days?*

(A) $\frac{11}{72}$ (B) $\frac{5}{18}$ (C) $\frac{1}{4}$ (D) $\frac{1}{2}$ (E) $\frac{2}{3}$

Problem 2.263. *Positive integer n is called "amazing", if the sum of its two greatest divisors is equal to 42. What is the total number of all "amazing" numbers?*

(A) 4 (B) 3 (C) 5 (D) 2 (E) 6

Problem 2.264. *Let $ABCDEF$ be an inscribed hexagon. Given that the measures (in degrees) of angles A, B, C, D, E form an increasing arithmetic sequence (in this order) with a difference of $10°$. What is the measure (in degrees) of angle A?*

(A) 90 (B) 100 (C) 120 (D) 135 (E) 150

Problem 2.265. *At most how many three-digit numbers can have the same sum of the digits?*

(A) 62 (B) 69 (C) 70 (D) 80 (E) 120

Problem 2.266. *What is the value of the sum of the digits of the smallest positive integer n, where $n \neq 1$, such that n and n^3 leave the same remainder after division by 2020^2?*

(A) 2 (B) 3 (C) 4 (D) 6 (E) 13

Problem 2.267. *Let AD be the bisector of angle BAC in triangle ABC. Given that $BD = 3$, $CD = 5$ and $AC - AB = 4$. What is the measure (in degrees) of $\angle ABC$?*

(A) 120 (B) 100 (C) 90 (D) 75 (E) 60

Problem 2.268. *What is the total number of all five-digit numbers divisible by 13 and ending with 12?*

(A) 23 (B) 27 (C) 60 (D) 68 (E) 69

Problem 2.269. *Let E and F be respectively the midpoints of sides AB and BC of square $ABCD$. Line segment EC intersects line segments AF, DF at points P, K respectively. Given that $AB = 5\sqrt{6}$. What is the area of triangle FPK?*

(A) $\sqrt{6}$ (B) 3 (C) 4 (D) $2\sqrt{6}$ (E) 5

Problem 2.270. What is the total number of all positive divisors of the number 10! which are multiples of 3?

(A) 100   (B) 200   (C) 216   (D) 300   (E) 420

Problem 2.271. A three-digit number $\overline{abc}$ written using some of the digits 0, 1, 2, 3, 4 is randomly chosen. What is the probability that the inequalities $|a - b| \geq 2$ and $|b - c| \geq 2$ simultaneously hold true?

(A) 0.05   (B) 0.1   (C) 0.2   (D) 0.23   (E) 0.25

Problem 2.272. At most how many pairwise distinct integers are there, such that the positive difference of any two of them is a prime number?

(A) 3   (B) 4   (C) 5   (D) 6   (E) 10

Problem 2.273. Let integer a be the smallest value of variable x, such that the values of the functions $y = \dfrac{2x}{3} - \dfrac{1}{3}$ and $y = \dfrac{3x^2}{14} + \dfrac{x}{7} - \dfrac{1}{2}$ are positive integers. What is the value of the sum of the digits of the number a^2?

(A) 4   (B) 5   (C) 6   (D) 7   (E) 10

Problem 2.274. Let $A_1B_1C_1$ be a triangle, such that $A_1B_1 = 4$, $A_1C_1 = 5$ and $B_1C_1 = 7$. Consider the sequence of triangles $A_nB_nC_1$ constructed as follows: for any positive integer n points B_{n+1} and A_{n+1} lie on sides C_1A_n and C_1B_n, respectively. Given that $\angle C_1B_{n+1}A_{n+1} = \angle C_1B_nA_n$ and that $A_nB_{n+1}A_{n+1}B_n$ is a tangential quadrilateral. What is the value of the sum of the perimeters of all triangles $A_nB_nC_1$?

(A) 18   (B) 20   (C) 24   (D) 32   (E) 60

Problem 2.275. A permutation $a_1, a_2, ..., a_8$ of the numbers 1, 2, ..., 8 is called "charming" if for any number $i \in \{1, 2, ..., 8\}$ either $i \mid a_i$ or $\lfloor \frac{i}{2} \rfloor \mid a_i$. What is the total number of all "charming" permutations of numbers 1,2,...,8?

(A) 252   (B) 216   (C) 192   (D) 120   (E) 96

2.12 AMC 10 type practice test 12

Problem 2.276. *What is the value of the expression* $(15^{-2} + 20^{-2} - 12^{-2})(15^2 + 20^2 - 12^2)$ *?*

(A) $\dfrac{1}{20}$ (B) $\dfrac{1}{12}$ (C) 1 (D) 0 (E) -1

Problem 2.277. *At first, David and Anna ate 7 candies together, then David ate another 5 candies. In the end it turned out that David ate twice as many candies as Anna. How many candies did Anna eat?*

(A) 3 (B) 4 (C) 5 (D) 6 (E) 2

Problem 2.278. *Given two positive numbers, such that the first number is greater than the second number by 10%. Given also that the difference of these two numbers is equal to 3. What is the value of the first number?*

(A) 33 (B) 30 (C) 27 (D) 25 (E) 12

Problem 2.279. *Five two-digit numbers are formed with the digits 0, 1, 2, 3, 4, 5, 6, 7, 8, 9, using each digit once. What is the possible smallest value of the sum of these five two-digit numbers?*

(A) 180 (B) 135 (C) 120 (D) 100 (E) 99

Problem 2.280. *Let a, b, c be real numbers, such that $ab - bc < 0$, $bc - ac < 0$ and $ab + ac < 0$. Which of the following conditions can hold true?*

(A) $a > 0, b > 0, c < 0$ (B) $a > 0, b > 0, c > 0$ (C) $a < 0, b < 0, c > 0$ (D) $a < 0, b > 0, c < 0$ (E) $a < 0, b > 0, c > 0$

Problem 2.281. *Given that today is not Wednesday. What is the probability of the event that tomorrow is Wednesday?*

(A) 0 (B) $\dfrac{1}{7}$ (C) $\dfrac{1}{6}$ (D) $\dfrac{2}{7}$ (E) $\dfrac{3}{7}$

Problem 2.282. *The sum of the ages of all classmates is 120. In 6 years the sum of theirs ages will be twice the sum of their ages 3 years ago. What is the value of the arithmetic mean of their ages?*

(A) 8 (B) 10 (C) 12 (D) 15 (E) 16

Problem 2.283. *Let $ABCD$ be a convex quadrilateral, such that $AD = 2BC$, $BD = BC$, $\angle ABC = 130°$ and $\angle BCD = 70°$. What is the measure (in degrees) of angle ADC?*

(A) 60 (B) 90 (C) 120 (D) 125 (E) 130

Problem 2.284. *David solved 5 problems each Saturdays and Sundays, on each week-day he solved 6 problems. Given that in a few consecutive days David solved 70 problems. Which day of the week did he start to solve these problems?*

(A) Wednesday (B) Thursday (C) Friday (D) Sunday (E) Monday

Problem 2.285. *Let $ABCD$ be a square with a side length $\sqrt{3}$. Let square $ABCD$ be rotated around point A by $30°$ and as a result we obtained square $AB'C'D'$. What is the area of the part that squares $AB'C'D'$ and $ABCD$ have in common?*

(A) 1 (B) 1.5 (C) $\sqrt{3}$ (D) 2 (E) 2.1

Problem 2.286. *Given that the intersection points of the graphs of the functions $y = x^2 + ax + 1$ and $y = -x^2 - 3x + b$ are symmetric with respect to point $(-1, 5)$. What is the value of $a + b$?*

(A) 10   (B) 8   (C) 7   (D) 5   (E) -6

Problem 2.287. *Given three pairwise different increasing arithmetic sequences, each consisting of one hundred integer terms. Let n be the total number of all those terms, each of which simultaneously belongs to these three sequences. What is the greatest possible value of n?*

(A) 50   (B) 34   (C) 33   (D) 25   (E) 16

Problem 2.288. *Three sisters bought 4 identical bracelets. In how many different ways can they wear those 4 bracelets? (They can wear each bracelet either on right or left hand).*

(A) 15   (B) 20   (C) 24   (D) 120   (E) 126

Problem 2.289. *What is the greatest possible number of all distinct prime numbers, such that the positive difference of any two of them is also a prime number?*

(A) 2   (B) 3   (C) 4   (D) 5   (E) 10

Problem 2.290. *Let $\{a\} = a - \lfloor a \rfloor$, where $\lfloor a \rfloor$ is the greatest positive integer not greater than a. What is the total number of the solutions of the equation $x + \{x\}^2 = 2020$?*

(A) 0   (B) 1   (C) 2   (D) 3   (E) 2020

Problem 2.291. *Jack shoots at a moving circular disk that rotates around its center. The probability to score 1 point, 2 points, 3 points with one shot is $\dfrac{1}{2}, \dfrac{1}{3}, \dfrac{1}{6}$, respectively. What is the probability of the event that Jack scores 6 points with three shots?*

(A) $\dfrac{1}{7}$ (B) $\dfrac{11}{54}$ (C) $\dfrac{1}{6}$ (D) $\dfrac{1}{3}$ (E) $\dfrac{1}{2}$

Problem 2.292. *Given that the sum of three smallest positive divisors of a positive integer n is equal to 7 and the sum of three greatest divisors of n is equal to 84. What is the value of the sum of all digits of n?*

(A) 8   (B) 9   (C) 10   (D) 11   (E) 12

Problem 2.293. *Let θ be the circumscribed circle of triangle ABC and $\angle A - \angle B = 50°$. Let the perpendicular bisector of line segment AB intersects the arc ACB of circle θ at point M. What is the measure (in degrees) of $\angle MAC$?*

(A) 20   (B) 25   (C) 30   (D) 15   (E) 10

Problem 2.294. *What is the value of the sum of all integers x, such that the the double inequality $|x+1| \leq |x-27| \leq |x+3|$ holds true?*

(A) 15   (B) 21   (C) 23   (D) 25   (E) 30

Problem 2.295. *Let $ABCD$ be a square with a side lenght of $2\sqrt{5}$. Let M, N, P, K be the midpoints of sides AB, BC, CD, AD, respectively. What is the radius of the circle that is tangent to each of line segments AN, DN, AP, BP, BK, CK, CM, DM?*

(A) $\dfrac{\sqrt{5}}{2}$ (B) 1   (C) $\dfrac{\sqrt{3}}{2}$ (D) $\dfrac{\sqrt{2}}{2}$ (E) $\dfrac{1}{2}$

Problem 2.296. Given 25 chairs around a round table, such that on each chair exactly one person is sitting. After the break each of them can sit on the fifth chair that comes after the chair he/she was sitting before the break (toward either direction). The counting is done starting from the next chair. On each chair is again sitting exactly one person. After the break, in how many different ways can these 25 people sit on these 25 chairs?

(A) 1024 (B) 32 (C) $(5!)^5$ (D) 5! (E) 120

Problem 2.297. Given a triangle ABC, such that $AB < BC$. Let D be the midpoint of arc AC of the circumcircle of triangle ABC, such that points B and D are on different sides of the line AC. Let segment DE be perpendicular to chord BC. Given that $BE = 17$ and $EC = 7$. What is the length of side AB?

(A) 7 (B) 8 (C) 9 (D) 10 (E) 11

Problem 2.298. Given that the angle measures of each of $\dfrac{n^2}{100}$ angles of a convex $n-$gon is equal to n. What is the total number of all possible values of n?

(A) 0 (B) 1 (C) 10 (D) 11 (E) infinitely many

Problem 2.299. Positive integer n is called "strange" if numbers $d(n)$, $d(n^2)$, $d(n^3)$ form an airthmetic sequence, where $d(m)$ is the total number of all positive integer divisors of a positive integer m. What is the total number of all two-digit "strange" numbers?

(A) 10 (B) 12 (C) 21 (D) 28 (E) 32

Problem 2.300. Let n be a positive integer greater than 3. Given that numbers $1, 2,..., n$ can be divided into several groups, such that the greatest element of each group is equal to the sum of all the other elements of that group. What is the smallest possible value of n?

(A) 10 (B) 11 (C) 12 (D) 13 (E) 14

Chapter 3

Answers

3.1 Answers of AMC 10 type practice tests

Problem	Practice test 1	Practice test 2	Practice test 3	Practice test 4
1	D	C	B	B
2	B	B	D	A
3	A	D	B	D
4	C	A	C	C
5	E	E	D	E
6	B	B	B	B
7	D	E	E	B
8	B	D	B	D
9	D	C	B	C
10	B	A	C	D
11	A	B	D	E
12	E	E	E	A
13	D	D	B	B
14	B	C	C	A
15	A	A	A	C
16	C	E	D	E
17	A	B	B	E
18	D	D	E	D
19	B	E	D	B
20	A	C	C	A
21	C	A	A	D
22	E	D	E	C
23	D	B	D	E
24	E	E	B	B
25	C	C	C	A

Problem	Practice test 5	Practice test 6	Practice test 7	Practice test 8
1	B	C	B	C
2	D	B	C	A
3	A	D	D	B
4	E	E	A	D
5	C	A	A	A
6	B	E	B	E
7	A	E	A	E
8	D	B	E	B
9	E	A	E	C
10	C	E	C	C
11	B	D	A	B
12	C	B	C	B
13	A	B	D	E
14	D	A	A	C
15	B	C	D	B
16	E	E	A	A
17	E	B	A	C
18	D	D	C	D
19	A	B	B	C
20	C	C	C	E
21	E	A	C	E
22	E	B	B	A
23	A	D	D	B
24	C	C	E	B
25	D	E	B	A

Problem	Practice test 9	Practice test 10	Practice test 11	Practice test 12
1	E	B	B	D
2	C	C	E	B
3	E	E	A	A
4	C	D	D	A
5	B	B	B	D
6	E	C	E	C
7	D	D	D	C
8	D	C	E	E
9	E	A	E	E
10	C	A	E	C
11	A	B	B	B
12	D	D	A	B
13	D	D	B	E
14	A	E	B	B
15	D	D	C	C
16	B	A	C	B
17	C	C	C	E
18	C	C	E	B
19	C	E	E	D
20	D	B	C	B
21	E	A	D	B
22	D	D	B	D
23	D	D	A	B
24	E	E	D	D
25	E	B	A	C

Chapter 4

Solutions

4.1 Solutions of AMC 10 type practice test 1

Problem 4.1. *A train left City A at 8 : 00 AM and arrived to City B after 45 minutes. It stopped in City B for 10 minutes and continued to City C. The train covered the distance between City B and City C in 75 minutes. At what time did the train arrive to City C?*

(A) 11:00 AM   (B) 10:05 AM   (C) 10:00 AM   (D) 10:10 AM   (E) 10:30 AM

Solution. Answer. (D)
The train covered the distance between cities A and C in $45 + 10 + 75 = 130$ minutes or 2 hours and 10 minutes. Therefore, it reached city C at 10 : 10 AM. □

Problem 4.2. *Let M and N be the midpoints of sides BC and CD of rectangle ABCD, respectively. The area of triangle AMN is what percent of the area of rectangle ABCD?*

(A) 25   (B) 37.5   (C) 30   (D) 24   (E) 50

Solution. Answer. (B)

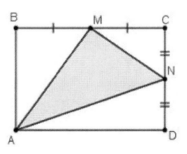

Let $AB = a, AD = b$. Thus, it follows that
$$BM = MC = \frac{b}{2}, CN = ND = \frac{a}{2}.$$

Therefore
$$Area(\triangle ABM) = \frac{ab}{4}, Area(\triangle ADN) = \frac{ab}{4}, Area(\triangle CMN) = \frac{ab}{8}.$$

Hence
$$Area(\triangle AMN) = ab - \frac{ab}{4} - \frac{ab}{4} - \frac{ab}{8} = \frac{3ab}{4}.$$

Therefore
$$\frac{Area(\triangle AMN)}{Area(ABCD)} = \frac{3}{8}.$$

That is 37.5%. □

Problem 4.3. *Let $C(n)$ be the sum of all distinct prime divisors of a positive integer n, where $n > 1$. What is $C(C(C(2008)))$?*

(A) 19   (B) 34   (C) 253   (D) 30   (E) 20

Solution. Answer. (A)
We have $2008 = 2^3 \cdot 251$. Thus, it follows that $C(2008) = 2 + 251 = 253 = 11 \cdot 23$. Therefore $C(C(2008)) = C(253) = 34$. Hence $C(34) = 2 + 17 = 19$. □

Problem 4.4. *Six painters working at the same constant rate, can completely paint an apartment in 8 hours. How many painters were working if it took 12 hours to paint that apartment?*

(A) 5   (B) 3   (C) 4   (D) 2   (E) 1

Solution. Answer. (C)
As six painters are painting (at the same rate) the apartment in 8 hours, thus one painter can paint the same apartment in 48 hours. Therefore, four painters can paint it in 12 hours. □

Problem 4.5. *What is the value of the expression* $\dfrac{1^2 + 1 \cdot 2 + 2^2}{1^3 \cdot 2^3} + \dfrac{2^2 + 2 \cdot 3 + 3^2}{2^3 \cdot 3^3} + ... + \dfrac{9^2 + 9 \cdot 10 + 10^2}{9^3 \cdot 10^3}$?

(A) 1   (B) 0.9   (C) 0.99   (D) 0.5   (E) 0.999

Solution. Answer. (E)
Note that
$$\frac{k^2 + k \cdot (k+1) + (k+1)^2}{k^3 \cdot (k+1)^3} = \frac{k^2 + k(k+1) + (k+1)^2}{k^3 \cdot (k+1)^3} = \frac{(k+1-k) \cdot (k^2 + k(k+1) + (k+1)^2)}{k^3 \cdot (k+1)^3} =$$
$$= \frac{(k+1)^3 - k^3}{k^3 \cdot (k+1)^3} = \frac{1}{k^3} - \frac{1}{(k+1)^3},$$

for $k = 1, 2, ..., 9$. Therefore, the given sum is equal to
$$1 - \frac{1}{2^3} + \frac{1}{2^3} - \frac{1}{3^3} + ... + \frac{1}{9^3} - \frac{1}{10^3} = 1 - \frac{1}{10^3} = 0.999.$$
□

Problem 4.6. *The path from A to B consists of a 10 kilometers stretch of flat land, 12 kilometers uphill and 16 kilometers downhill. A car covered the respective stretches at speeds of 80 kmh, 48 kmh, and 90 kmh. What was car's average speed in kmh (to the nearest integer number) while traveling from A to B?*

(A) 68   (B) 69   (C) 70   (D) 72   (E) 60

Solution. Answer. (B)
The car has spent
$$\frac{10}{80} + \frac{12}{48} + \frac{16}{90} = \frac{199}{360}$$
hours on the "AB-road" of distance $10 + 12 + 16 = 38$ km. Thus, it follows that its average speed is equal to
$$\frac{38}{\frac{199}{360}} = 68\frac{148}{199} \approx 68.7 \; kph.$$
□

Problem 4.7. What is the value of the expression $\dfrac{(2^{2008})^2 + 2^{2007}}{(2^{2007})^2 + 2^{2005}}$?

(A) 2^{2007} (B) 2^{2008} (C) 9 (D) 4 (E) 5

Solution. Answer. (D)
We have that
$$\frac{(2^{2008})^2 + 2^{2007}}{(2^{2007})^2 + 2^{2005}} = \frac{2^{2007}(2^{2009} + 1)}{2^{2005}(2^{2009} + 1)} = 2^2 = 4.$$

$\square$

Problem 4.8. *A shop bought a coat for $500. The shop inteded to sell the coat for some price, but sold it 5% more than the intended sales price. Given that for that deal the shop generated a total profit of $67. What was the intended sales price of the coat?*

(A) $560 (B) $540 (C) $550 (D) $530 (E) $520

Solution. Answer. (B)
Let x be the intended sales price of the coat. Thus, it follows that
$$1.05x - 67 = 500.$$
Therefore $x = 540$.

$\square$

Problem 4.9. *For how many positive values of x is $\dfrac{20}{x+1} + \dfrac{1}{3(x+1)}$ a positive integer?*

(A) 0 (B) 10 (C) 19 (D) 20 (E) 25

Solution. Answer. (D)
Note that
$$\frac{20}{x+1} + \frac{1}{3(x+1)} = \frac{61}{3(x+1)} < \frac{61}{3}.$$
Therefore, the fraction
$$\frac{61}{3(x+1)}$$
can attain natural values from 1 to 20. Let $n = \{1, 2, ..., 20\}$ and
$$\frac{61}{3(x+1)} = n.$$
Thus, it follows that
$$x = \frac{61 - 3n}{3n} > 0.$$

$\square$

Problem 4.10. Let S_1 be a square with a side length 7. The vertices of square S_2 are located on the sides of S_1. The sides of square S_3 are parallel to the sides of S_1, and the vertices of square S_3 are located on the sides of square S_2. Given that the side length of square S_3 is equal to 4. What is the value of the side length of square S_2?

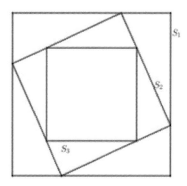

(A) $\sqrt{13}$ (B) $2\sqrt{7}$ (C) 6   (D) 5   (E) $4\sqrt{2}$

Solution. Answer. (B)

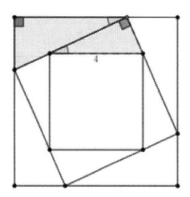

Let us denote by x the side of square S_2. Note that the shaded right-angle triangles are similar. Therefore, the ratio of their perimeters is the same as the ratio of respective hypotenuses. Thus, it follows that
$$\frac{x+7}{x+4} = \frac{x}{4}.$$
Hence, we obtain that $x = 2\sqrt{7}$. □

Problem 4.11. *A fisherman makes a round-trip and takes his boat 10 miles into a lake from its shore and back (covering the same distance). The average speed of the boat is 2.5 miles per hour. The fisherman catches (in average) 0.5 pounds of fish per hour. How many pounds of fish does the fisherman catch per one round-trip?*

(A) 4   (B) 3   (C) 2   (D) 5   (E) 6

Solution. Answer. (A)

Note that the fisherman boats in the lake $\dfrac{20}{2.5} = 8$ hours. So, one-round trip takes 8 hours.
Given that the fisherman catches (in average) 0.5 pounds of fish per hour.
Therefore, the fisherman catches $8 \cdot 0.5 = 4$ pounds of fish per one round-trip. □

Problem 4.12. *In one school there are 25% less 11th graders than 10th graders, and 20% more 11th graders than 12th graders. The total number of students in 10th, 11th, and 12th grades in that school is equal to 190. How many 10th graders are there in that school?*

(A) 60   (B) 50   (C) 40   (D) 100   (E) 80

Solution. Answer. (E)
Let us denote by x the number of 11th grade students. Therefore, the number of 10th graders is equal to $\frac{4}{3}x$ and the number of 12th grade students is equal to $\frac{5}{6}x$. Thus, according to the given condition we have that
$$x + \frac{4}{3}x + \frac{5}{6}x = 190.$$
Hence, we obtain that $x = 60$. Therefore
$$\frac{4}{3}x = 80.$$

□

Problem 4.13. *One combine harvester can harvest a field on its own in 12 hours. A second combine harvester can harvest the same field on its own in 18 hours. How many hours will it take both combine harvesters, working together, to harvest the whole field, if they also take a 48 minute break?*

(A) 7.2   (B) 7   (C) 9   (D) 8   (E) 10

Solution. Answer. (D)
First combine harvester's rate is $\frac{1}{12}$ of the field per hour, while the same for the second combine harvester is $\frac{1}{18}$ per hour. Thus, both together can harvest $\frac{1}{12} + \frac{1}{18} = \frac{5}{36}$ of the field per hour. Therefore, it would take them $\frac{36}{5} = 7.2$ hours to complete the job and if they take rest for 48 min (or 0.8 hours), then they can complete the work in $7.2 + 0.8 = 8$ hours. □

Problem 4.14. *The width and length of the frame of a painting are in the proportion 4:5. The respective dimensions of the painting are in the proportion 3:4 and the painting is the same distance from the frame on each side. What is the ratio of the area of the frame to the area of the painting?*

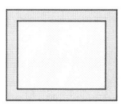

(A) $\frac{1}{2}$ (B) $\frac{2}{3}$ (C) $\frac{1}{3}$ (D) $\frac{1}{4}$ (E) $\frac{1}{10}$

Solution. Answer. (B)

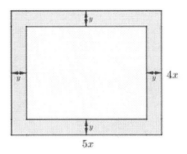

Let the width of the frame be $4x$ and the length be $5x$. Given that

$$\frac{5x - 2y}{4x - 2y} = \frac{4}{3}.$$

Thus, it follows that $x = 2y$. Therefore, the ratio of the area of the frame to the area of the painting is:

$$\frac{20x^2 - (5x - 2y)(4x - 2y)}{(5x - 2y)(4x - 2y)} = \frac{80y^2 - 48y^2}{48y^2} = \frac{2}{3}.$$

Alternative solution. Let the width of the painting be $3z$ and the length be $4z$. Thus, it follows that

$$5x - 4z = 4x - 3z.$$

Hence, we obtain that $x = z$. Therefore, we can take $x = z = 1$. Thus, the area of the painting is 12sq. units and the area of the frame is $5 \cdot 4 - 3 \cdot 4 = 8$ sq. units. Hence, the ratio of the area of the frame to the area of the painting is:

$$\frac{8}{12} = \frac{2}{3}.$$

$\square$

Problem 4.15. *A boat covers the distance between piers A and B traveling downstream in two less hours than it covers the same distance traveling upstream. A raft takes 6 more hours to cover the distance from pier A to B than the boat takes to cover the same distance traveling downstream. How many hours does it take the raft to cover the distance between piers A and B?*

(A) 8   (B) 9   (C) 9.5   (D) 10   (E) 10.5

Solution. Answer. (A)
Let boat's own speed be x kmh and the speed of river's flow be y kmh. If the raft covers the distance S km between the piers A and B in t hours, then

$$ty = (t - 6)(x + y) = (t - 4)(x - y).$$

Thus, it follows that

$$\frac{x + y}{y} = \frac{t}{t - 6},$$

and

$$\frac{x - y}{y} = \frac{t}{t - 4}.$$

As, we have that

$$2 = \frac{x + y}{y} - \frac{x - y}{y} = \frac{t}{t - 6} - \frac{t}{t - 4}.$$

Hence, we obtain that $(t - 6)(t - 4) = t$ or $t = 8$. $\square$

Problem 4.16. Let AB be the diameter of a semicircle with center O. Given semicircles with diameters OA, OB and circle S, such that S is tangent to these three semicircles (see the figure). Let r be the radius of circle S. What is the value of $\dfrac{AB}{r}$?

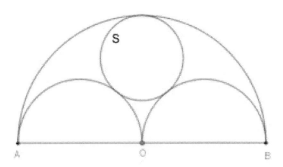

(A) 5   (B) 4   (C) 6   (D) 4.5   (E) 3

Solution. Answer. (C)

Let us denote the centers of semicircles (with diameters OA and OB) by O_1 and O_2, respectively, and the center of circle S by O_3. Let $OA = R$, then

$$O_2O_3 = O_1O_2 = \frac{R}{2} + r, \quad OO_3 = R - r,$$

and

$$OO_1 = OO_2 = \frac{R}{2}.$$

Therefore $OO_3 \perp O_1O_2$. Thus, from the right-angle $\triangle OO_1O_3$ and by Pythagorean theorem we have that

$$\left(\frac{R}{2} + r\right)^2 = \left(\frac{R}{2}\right)^2 + (R - r)^2.$$

Hence, we obtain that $R = 3r$ or

$$\frac{AB}{r} = \frac{2R}{r} = 6.$$

□

Problem 4.17. *Given a triangle with area of 10 sq. units and an inradius of 2 units. What is the area of the figure that consists of all points on the plane that are not more than 1 unit away from the sides of this triangle?*

(A) $20 + \pi$ (B) 20   (C) $20 + 2\pi$ (D) $10 + \pi$ (E) $17.5 + \pi$

Solution. Answer. (E)

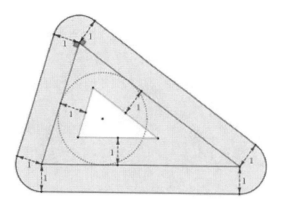

It is not difficult to note that the required figure is the shaded region shown. At first, let us find the area of the triangle inside the shaded region. From the similarity of the initial and smaller triangles and due to the fact that distance between the sides is equal to the half of the inradius, we can state that sides of the smaller triangle are equal to the half of the respective sides of the initial triangle. Therefore, its area is $\left(\frac{1}{2}\right)2 \cdot 10 = 2.5$ sq. units. Thus, it follows that the area of the part of the shaded region between the triangles is equal to $10 - 2.5 = 7.5$ sq. units. Now, note that the area of the part of the shaded region outside the initial triangle is $P \cdot 1 + \pi \cdot 1^2$ sq. units, as the three circular sectors together make a complete circle with radius 1 unit and the three rectangular areas make a rectangle with "length" equal to the perimeter of the initial triangle and "width" eqaul to 1 unit. Therefore, the perimeter of the initial triangle is equal to:

$$P = \frac{2S}{r} = \frac{2 \cdot 10}{2} = 10.$$

Thus, the area of the shaded region is $7.5 + (10 + \pi) = 17.5 + \pi$ square units. □

Problem 4.18. *Let I be the incenter of right triangle $\triangle ABC$ with hypotenuse AB. What is the ratio of the circumradius of $\triangle ABC$ to the circumradius of $\triangle ABI$?*

(A) 2 (B) $\frac{2}{3}$ (C) $\frac{1}{2}$ (D) $\frac{\sqrt{2}}{2}$ (E) $\sqrt{2}$

Solution. Answer. (D)

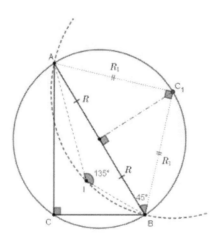

We have that

$$\angle AIB = 180° - \angle IAB - \angle IBA = 180° - \frac{1}{2}\angle A - \frac{1}{2}\angle B = 180° - \frac{1}{2}(\angle A + \angle B) = 135°.$$

According to the law of sines and denoting by R and R_1 the curcumradii of $\triangle ABC$ and $\triangle ABI$, respectively, we obtain that
$$R = \frac{AB}{2 \cdot \sin 90°},$$
and
$$R_1 = \frac{AB}{2 \cdot \sin 135°}.$$
Thus, it follows that
$$\frac{R}{R_1} = \frac{\sin 135°}{\sin 90°} = \frac{\sqrt{2}}{2}.$$

Alternative solution. Let C_1 be the circumcenter of $\triangle ABI$. As $\angle AIB = 135°$, then
$$\angle AC_1B = 360° - 2 \cdot 135° = 90°.$$

Therefore, the points A, B, C and C_1 lie on the same circle (i.e. $ABCC_1$ is a cyclic quadrilateral) and $\triangle AC_1B$ is an isosceles right triangle with hypotenuse AB and legs equal to R_1. Thus, it follows that $R_1 = R_2$. □

Problem 4.19. *Let A be a point on circle $\triangle$ with center C and radius 5. Circle $\triangle$ is rotated 30° counterclockwise around point A. What is the length of the arc of the second circle which is inside of the first circle?*

(A) $\dfrac{35\pi}{6}$ (B) $\dfrac{25\pi}{6}$ (C) 4π (D) 3π (E) 5π

Solution. Answer. (B)

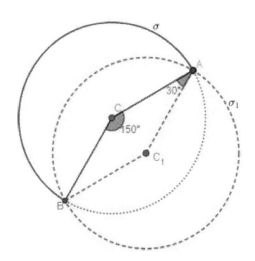

Let $\triangle_1$ be the rotated circle (30° counterclockwise around point A) of circle $\triangle$. Let the intersection points between $\triangle$ and $\triangle_1$ be the points A and B as shown in the figure. We have that
$$AC = BC = AC_1 = BC_1 = 5.$$

Thus, it follows that $ACBC_1$ is a rhombus with acute angle $\angle CAC_1 = 30°$. Therefore, central angle $\angle ACB = 150°$. Hence, the smaller arclength of AB is equal to
$$\frac{2\pi \cdot 5}{360} \cdot 150 = \frac{25\pi}{6}.$$

□

Problem 4.20. *Let the diagonals of convex quadrilateral $ABCD$ intersect at point O. Given that the areas of $\triangle ABO$, $\triangle CBO$, $\triangle COD$ are equal to 6, 9, 18, respectively. What is the area of quadrilateral $ABCD$?*

(A) 45 (B) 50 (C) 36 (D) 30 (E) 24

Solution. Answer. (A)

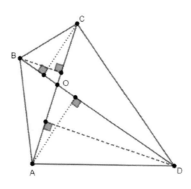

Note that $\triangle ABO$ and $\triangle CBO$ share the same height. Therefore
$$\frac{Area(ABO)}{Area(CBO)} = \frac{6}{9} = \frac{AO}{OC}.$$

Similarly, we have that
$$\frac{Area(AOD)}{Area(COD)} = \frac{AO}{OC}.$$

It follows that
$$\frac{6}{9} = \frac{Area(AOD)}{18}.$$

Thus, $Area(AOD) = 12$.
Hence, $Area(ABCD) = 6 + 9 + 18 + 12 = 45$. □

Problem 4.21. *Let $ABCDA_1B_1C_1D_1$ be a cube with a side length of 2. Let M and N be points on sides B_1C_1 and C_1D_1, such that $BMND$ is a tangential quadrilateral. What is the length of line segment BM?*

(A) $6\sqrt{2} - 4\sqrt{3}$ (B) $\sqrt{5}$ (C) $4\sqrt{2} - 2\sqrt{3}$ (D) $\dfrac{\sqrt{17}}{2}$ (E) $\sqrt{2} + \sqrt{3}$

Solution. Answer. (C)

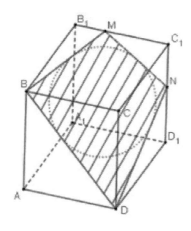

Let $B_1M = x$. Thus, it follows that $C_1M = 2 - x$. As $MN \parallel B_1D_1$, then
$$C_1N = C_1M = 2 - x.$$

From $\triangle BB_1M$ and $\triangle MC_1N$ and from Pythagorean theorem, it follows that
$$BM = \sqrt{4 + x^2},$$
and
$$MN = \sqrt{2}(2 - x).$$

As $BMND$ is a tangential quadrilateral, then
$$BM + ND = BD + MN.$$

Thus, it follows that
$$2\sqrt{4 + x^2} = 2\sqrt{2} + \sqrt{2}(2 - x).$$

Therefore
$$2(4 + x^2) = (4 - x)^2.$$

We deduce that
$$x^2 + 8x = 8.$$

Thus, it follows that
$$x = 2\sqrt{6} - 4.$$

Hence, we obtain that
$$BM = \frac{1}{2}(BD + MN) = \frac{\sqrt{2}}{2}(4 - x) = \frac{\sqrt{2}}{2}(8 - 2\sqrt{6}) = 4\sqrt{2} - 2\sqrt{3}.$$

$\square$

Problem 4.22. *Given 36 different rectangles of size $m \times n$, where $m \leq n \leq 8$ and m, n are positive integers. What is the probability that if two rectangles are randomly chosen, then neither of them can be covered with the other one such that the corresponding sides of the rectangles are parallel?*

(A) $\dfrac{6}{35}$ (B) $\dfrac{1}{63}$ (C) $\dfrac{11}{630}$ (D) $\dfrac{1}{2}$ (E) $\dfrac{1}{5}$

Solution. Answer. (E)

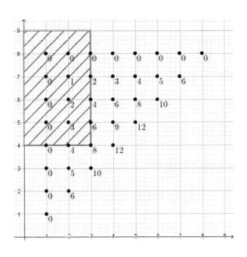

Let us place the points (m, n) on the xy−coordinate plane instead of $m \times n$ rectangles. So, 36 rectangles are replaced with 36 points. Let us write next to each point number of points located "north-west" of that point. So, total number of outcomes will be $\binom{36}{2} = 18 \cdot 35$. While the number of favorable outcomes will be the sum of all numbers next to each point: $1 + 4 + 10 + 20 + 35 + 56 = 126$. Thus, the required probability is equal to
$$\frac{126}{18 \cdot 35} = \frac{1}{5}.$$
$\square$

Problem 4.23. *Let S be the set of all positive integers from 1 to 20. How many subsets of S exist, such that each of them contains at least one prime number?*

(A) 4096   (B) 2^{19} (C) 4   (D) 1044480   (E) 1024

Solution. Answer. (D)
Let us first find the number of subsets of S such that each of them does not contain any prime number. Each of such subsets is a subset of $\{1, 4, 6, 8, 9, 10, 12, 14, 15, 16, 18, 20\}$. Therefore, there are 212 such subsets. Thus, the number of required subsets is
$$2^{20} - 2^{12} = 1044480.$$
$\square$

Problem 4.24. *What is the units digit of $2008^{2007^{2008}} + 2007^{2008^{2007}}$?*

(A) 1   (B) 3   (C) 5   (D) 7   (E) 9

Solution. Answer. (E)
Note that
$$4 \mid 2007^{2008} - 1 = (2007^2)^{1004} - 1^{1004}.$$
Thus, it follows that
$$2008^{2007^{2008}} = 2008^{4k+1},$$
where k is a positive integer. We have that 2008^4 ends with 6. Hence, 2008^{4k} also ends with 6. We have that
$$4 \mid 2008^{2007}.$$
Thus, it follows that
$$2007^{2008^{2007}} = 2007^{4m} = (2007^4)^m,$$
where m is a positive integer 2007^4 ends by 1. Hence $2007^{2008^{2007}}$ also ends by 1. Therefore, the required number ends by 9.
$\square$

Problem 4.25. *Let a square be constructed externally on each side of a regular hexagon with a side length of 1. What is the radius of the circle which passes through all the vertices of these squares that are not the vertices of the hexagon?*

(A) $\sqrt{3}$ (B) 2 (C) $\dfrac{\sqrt{6}+\sqrt{2}}{2}$ (D) $\sqrt{5}+1$ (E) $\sqrt{2}$

Solution. Answer. (C)

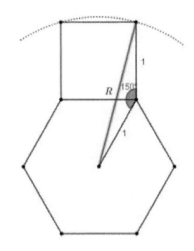

According to the law of cosines (see the figure), we have that
$$R^2 = 1^2 + 1^2 - 2\cdot 1\cdot 1\cdot \cos 150° = 2 + \sqrt{3}.$$
Thus, it follows that
$$R = \sqrt{2+\sqrt{3}} = \sqrt{\dfrac{4+2\sqrt{3}}{2}} = \dfrac{\sqrt{3}+1}{2} = \dfrac{\sqrt{6}+\sqrt{2}}{2}.$$

□

4.2 Solutions of AMC 10 type practice test 2

Problem 4.26. *A single ice cream costs $5. At most, how many ice creams can David buy with $63?*

(A) 13   (B) 11   (C) 12   (D) 10   (E) 9

Solution. Answer. (C)
$63 = 5 \cdot 12 + 3$, thus with 5\$ David can buy at most 12 ice-creams. □

Problem 4.27. *There is $500 in each of three envelopes. The first envelope contains only $10 bills, the second one-only $20 bills, and the third one-only $50 bills. One, two, and three bills are taken out of these envelopes (in any order). What is the smallest possible amount, in dollars, that can be taken out?*

(A) 130   (B) 120   (C) 110   (D) 230   (E) 100

Solution. Answer. (B)
The least possible amount is
$$50 \cdot 1 + 20 \cdot 2 + 10 \cdot 3 = 120.$$
□

Problem 4.28. *What is x, if $1 - \dfrac{1}{1 - \dfrac{1}{1+x}} = 2$?*

(A) -1   (B) 0   (C) $\dfrac{1}{3}$ (D) -0.5   (E) 0.5

Solution. Answer. (D)
We have that
$$1 - 2 = \frac{1}{1 - \frac{1}{1+x}}.$$
Thus, it follows that
$$1 - \frac{1}{1+x} = -1.$$
Hence, we obtain that
$$1 + x = \frac{1}{2}.$$
Therefore $x = -0.5$. □

Problem 4.29. *Ann's father drives her from home to school every day. The school is 20 miles away from home, and Ann noticed that it took them 20, 24, and 30 minutes in the last three days to reach school. What was their average speed during last three days (in miles per hour) rounded to the nearest whole number?*

(A) 49   (B) 50   (C) 60   (D) 55   (E) 45

Solution. Answer. (A)
Total distance travelled is equal to $3 \cdot 20 = 60$ miles. Total time travelled in hours is equal to
$$\frac{1}{3} + \frac{2}{5} + \frac{1}{2} = \frac{37}{30}.$$
Thus, the average speed in miles per hour is equal to
$$\frac{60}{\frac{37}{30}} = \frac{1800}{37} \approx 48.6.$$
□

Problem 4.30. *What is the value of the sum of the digits of* $200920092009 \cdot 999999999999$?

(A) 90   (B) 99   (C) 81   (D) 117   (E) 108

Solution. Answer. (E)
We have that

$$200920092009 \cdot 999999999999 - 200920092009 \cdot 1012 - 200920092009 = 200920092008799079907991.$$

Thus, it follows that the sum of the digits is equal to

$$11 \cdot 3 - 1 + 25 \cdot 3 + 1 = 36 \cdot 3 = 108.$$

$\square$

Problem 4.31. *A circle is inscribed into a circular sector with a central angle of* $90°$. *What is the ratio of the area of the circle to the area of the sector?*

(A) $\dfrac{4}{5}$ (B) $4(3 - 2\sqrt{2})$ (C) $\dfrac{2}{3}$ (D) $\dfrac{1}{3}$ (E) $4(\sqrt{2} - 1)$

Solution. Answer. (B)

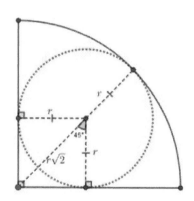

Let us denote by r the radius of the inscribed circle. Therefore, the radius of the sector is equal to

$$R = r\sqrt{2} + r.$$

Thus, it follows that the ratio of the areas of the circle to the sector is equal to

$$\frac{\pi r^2}{\frac{1}{4}\pi r^2(\sqrt{2} + 1)^2} = \frac{4}{(\sqrt{2} + 1)^2} = 4(3 - 2\sqrt{2}).$$

$\square$

Problem 4.32. *The weight of a fish's head is* 20% *of its total weight and 8 kilograms less than its total weight. How much (in kilograms) does the fish weigh?*

(A) 18   (B) 9   (C) 12   (D) 16   (E) 10

Solution. Answer. (E)
Let the total weight of the fish be x kilograms. Thus, it follows that its head (in kilograms) weighs

$$\frac{x \cdot 20}{100} = \frac{x}{5}.$$

On the other hand, it is given that
$$x - \frac{x}{5} = 8.$$
Therefore $x = 10$ kilograms. $\square$

Problem 4.33. *Eighty percent of the tourists in a group took a bus and the rest took a taxi. The bus fare per person is 40% less than the taxi fare per person. All together, they paid $68 for the tour transportation, and the bus fare per person is $3. How many tourists are in the group?*

(A) 15   (B) 16   (C) 18   (D) 20   (E) 10

Solution. Answer. (D)

Let the taxi fare be $x\$$ per person, therefore $x \cdot \frac{60}{100} = 3$. Thus, it follows that $x = 5$. Now, let the number of people in the group be y, then
$$\frac{y \cdot 80}{100} \cdot 3 + \frac{y \cdot 20}{100} \cdot 5 = 68.$$
Therefore
$$y = \frac{68 \cdot 5}{17} = 20.$$
$\square$

Problem 4.34. *Positive integers a, b, c and 27, where $a < b < c$, form a geometric sequence. What is the largest possible value of a?*

(A) 1   (B) 2   (C) 8   (D) 12   (E) 16

Solution. Answer. (C)

Let the common ratio of the geometric sequence be r, then
$$r = \frac{b}{a} = \frac{m}{n},$$
where the positive integers m and n are relatively prime, i.e. $gcd(m, n) = 1$. Therefore
$$a \left(\frac{m}{n} \right)^3 = 27.$$
Thus, it follows that
$$a = \left(\frac{3n}{m} \right)^3.$$
Hence, we obtain that $m \mid 3n$. We deduce that $a \leq 8$. Indeed, 8, 12, 18, 27 form a geometric sequence, thus the largest possible value of a is 8. $\square$

Problem 4.35. *In right triangle $\triangle ABC$, $\angle C = 90°$ and point H is the foot of the altitude drawn from vertex C. Given $AH = 3.6$ and $BH = 6.4$, what is the perimeter of $\triangle ABC$?*

(A) 24   (B) 12   (C) 18   (D) 36   (E) 16

Solution. Answer. (A)

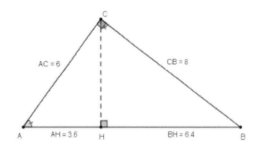

Note that $\triangle ACH$ is similar to $\triangle ABC$. Thus, it follows that
$$\frac{AC}{AB} = \frac{AH}{AC}.$$
Therefore
$$AC^2 = AH \cdot AB.$$
In a similar way, we obtain that
$$BC^2 = BH \cdot AB.$$
Therefore, the perimeter of $\triangle ABC$ is equal to
$$AC + BC + AB = \sqrt{3.6 \cdot 10} + \sqrt{6.4 \cdot 10} + 10 = 24.$$
$\square$

Problem 4.36. *If the width of a given rectangle is increased by 2 units and the length is decreased by 2 units, then the area would be 28sq. units. However, if the width of the original rectangle is decreased by 2 units and the length is increased by 2 units, then the area is 24sq. units. What is the area of the original rectangle?*

(A) 28   (B) 30   (C) 24   (D) 16   (E) 18

Solution. Answer. (B)
Let the width of the given rectangle be a and the length be b. It is given that
$$(a+2)(b-2) = 28,$$
and
$$(a-2)(b+2) = 24.$$
Summing up these two equations together, we obtain that $2ab - 8 = 52$. Therefore $ab = 30$. $\square$

Problem 4.37. *Let ABC be a triangle, such that $AB = 1$ and $BC = 12$. Given that the median drawn from vertex B is BM, whose length (in units) is a positive integer. What is the length of BM?*

(A) 3   (B) 4   (C) 7   (D) 5   (E) 6

Solution. Answer. (E)

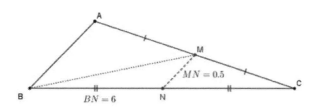

Let point N be the midpoint of the side BC. Thus, it follows that $MN = \frac{AB}{2} = 0.5$ and $BN = 6$. From $\triangle BMN$ and according to the triangle Inequality we obtain that $BN - NM < BM < BN + NM$. Therefore $5.5 < BM < 6.5$. Hence, we deduce that $BM = 6$. $\square$

Problem 4.38. *Let m and n be positive integers, such that $2^m \cdot 3^n = a$ and $2^n \cdot 3^m = b$. What is $3^{n^2-m^2}$?*

(A) $a^n \cdot b^m$ (B) $a^m \cdot b^{-n}$ (C) $(ab)^{n-m}$ (D) $a^{-n} \cdot b^{-m}$ (E) $a^n \cdot b^{-m}$

Solution. Answer. (D)
We have that
$$2^{mn} \cdot 3^{n^2} = a^n.$$
In a similar way, we obtain that
$$2^{mn} \cdot 3^{m^2} = b^m.$$
Therefore
$$\frac{a^n}{b^m} = 3^{n^2-m^2}.$$
Hence, we deduce that
$$3^{n^2-m^2} = a^n \cdot b^{-m}.$$
□

Problem 4.39. *Five unit squares are shown in the figure. What is the length of the side of the largest square?*

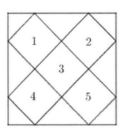

(A) 2   (B) 3   (C) $2\sqrt{2}$ (D) $\sqrt{7}$ (E) $\sqrt{6}$

Solution. Answer. (C)
Let the length of a side of the biggest square be equal to a units. Thus, it follows that its area is equal to
$$a^2 = 5 \cdot 1 + 4 \cdot \frac{1}{4} + 4 \cdot \frac{1}{2} = 8.$$
Therefore $a = 2\sqrt{2}$.   □

Problem 4.40. *What is the value of $1^2 + 2 \cdot 2^2 + 2 \cdot 3^2 + ... + 2 \cdot 19^2 + 20^2 + 1 \cdot 2 + 2 \cdot 3 + 3 \cdot 4 + ... + 19 \cdot 20$?*

(A) 7999   (B) 8000   (C) 7200   (D) 789   (E) 800

Solution. Answer. (A)
Let us rewrite the given sum in the following way
$$1^2 + 2 \cdot 2^2 + 2 \cdot 3^2 + ... + 2 \cdot 19^2 + 20^2 + 1 \cdot 2 + 2 \cdot 3 + 3 \cdot 4 + ... + 19 \cdot 20 =$$
$$= (2^2 + 2 \cdot 1 + 1^2) + (3^2 + 3 \cdot 2 + 2^2) + ... + (20^2 + 20 \cdot 19 + 19^2) = \frac{2^3 - 1^3}{2-1} + \frac{3^3 - 2^3}{3-2} + ... + \frac{20^3 - 19^3}{20-19} =$$
$$= 2^3 - 1^3 + 3^3 - 2^3 + ... + 20^3 - 19^3 = 20^3 - 1 = 7999.$$
□

Problem 4.41. *Points A_1, A_2, A_3, A_4, A_5 lie on the same plane. Given that the length of the line segments $A_1A_2 = 1, A_2A_3 = 2, A_3A_4 = 3, A_4A_5 = 4$. How many integer lengths are possible for line segment A_1A_5?*

(A) 10   (B) 9   (C) 6   (D) 5   (E) 11

Solution. Answer. (E)
According to the triangle inequality, we have that

$$A_1A_5 \leq A_1A_4 + A_4A_5, A_1A_4 \leq A_1A_3 + A_3A_4, A_1A_3 \leq A_1A_2 + A_2A_3.$$

Summing up these inequalities, we obtain that

$$A_1A_5 \leq A_1A_2 + A_2A_3 + A_3A_4 + A_4A_5 = 10.$$

Thus, it follows that
$$A_1A_5 \in \{0, 1, 2, 3, 4, 5, 6, 7, 8, 9, 10\}.$$

In the figures below, 11 possible integer lengths of A_1A_5 are shown:
Case 1.

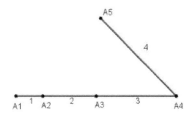

In Case 1, we have that $6 - 4 = 2 < A_1A_5 < 10 = 6 + 4$. Therefore $A_1A_5 \in \{3, 4, 5, 6, 7, 8, 9\}$.
Case 2.

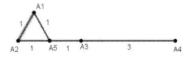

In Case 2, we have that $A_1A_5 = 0$.
Case 3.

In Case 3, we have that $A_1A_5 = 1$.
Case 4.

In Case 4, we have that $A_1A_5 = 2$.
Case 5.

In Case 5, we have that $A_1A_5 = 10$.   □

Problem 4.42. *In rectangle $ABCD$, $AB = 5$ and $BC = 12$. The line passing through vertex B and perpendicular to diagonal AC, intersects with side AD at point E. What is the length of line segment AE?*

(A) $1\dfrac{7}{12}$ (B) $2\dfrac{1}{12}$ (C) 2.4 (D) $2\dfrac{3}{5}$ (E) 2.8

Solution. Answer. (B)

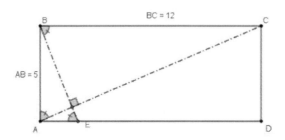

Note that
$$\angle AEB = \angle EBC = 90° - \angle BCA = \angle BAC.$$
Thus, it follows that $\triangle AEB$ and $\triangle ABC$ are similar. Therefore
$$\frac{AB}{BC} = \frac{AE}{AB}.$$
Hence, we deduce that
$$AE = \frac{AB^2}{BC} = \frac{25}{12} = 2\frac{1}{12}.$$

□

Problem 4.43. *Anna, Eric, and John bought some pens. If Anna gives 50% of her pens to Eric, and after that Eric gives 50% of his pens to John, and finally John gives $\dfrac{100}{3}$% of his pens to Anna, then all three would have an equal number of pens. Anna had more pens than John originally by what percent?*

(A) 50 (B) 60 (C) 40 (D) 100 (E) 80

Solution. Answer. (D)
Suppose they all had x pens at the end. Before John gave pens to Anna, John has $\dfrac{3x}{2}$ pens and Anna had $\dfrac{x}{2}$ pens, and Eric had x pens. Before that, John had $\dfrac{x}{2}$ pens, Eric $2x$, and Anna $\dfrac{x}{2}$ pens. Therefore, initially John had $\dfrac{x}{2}$ pens, Eric $\dfrac{3x}{2}$ pens, and Anna x pens. Thus, Anna had 100% more pens than John.

□

Problem 4.44. *Let a, b, c be positive integers that form a geometric sequence. Given that a has 3 divisors and c has 9 divisors. At most how many divisors can b have?*

(A) 14 (B) 9 (C) 6 (D) 8 (E) 10

Solution. Answer. (E)
Note that $a = p^2$, where p is some prime number. If $c_1, c_2, ..., c_9$ are all divisors of c, such that
$$c_1 < c_2 < ... < c_9.$$

Thus, it follows that
$$\frac{c}{c_9} < \frac{c}{c_8} < \ldots < \frac{c}{c_1}$$
are divisors of c. Therefore
$$\frac{c}{c_5} = c_5.$$

Hence, we deduce that $c = c_5{}^2$.

Therefore, the following cases are possible: $c_5 = p^4$ or $c_5 = q^4$ or $c_5 = pq$ or $c_5 = qr$, where $p, q,$ and r are pairwise relatively prime numbers. We obtain that $b = p^5$ or $b = pq^4$ or $b = p^2q$ or $b = pqr$. Thus, it follows that b can have at most $(1+1)(4+1) = 2 \cdot 5 = 10$ divisors. $\square$

Problem 4.45. *A jet ski leaves port B traveling upstream towards port A, and at the same time a raft starts drifting from port A towards port B; they meet each other 3 hours after their departure. If, simultaneously, another jet ski leaves port A travelling downstream, 5mph faster than the other jet ski, the two jet skis would meet 1 hour after their departure. What is the speed (in miles per hour) of the second jet ski?*

(A) 6   (B) 15   (C) 10   (D) 20   (E) 18

Solution. Answer. (C)

Let the speed of the first jet ski be v mph. Therefore, the speed of the second jet ski is $v + 5$ mph. Let the speed of the raft be r mph. Given that the distance between ports A and B is equal to $3(r + v - r)$. On the other hand, it is equal to
$$1 \cdot (v + 5 + r + (v - r)).$$
Hence, we deduce that
$$3v = 2v + 5.$$
Therefore $v = 5$. Thus, it follows that $v + 5 = 10$ mph. $\square$

Problem 4.46. *Three unit circles are tangent to diameter AB of the given semicircle (see the figure). Two of the unit circles are externally tangent to the third unit circle and the given semicircle. What is the length of diameter AB?*

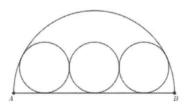

(A) $2 + 2\sqrt{5}$ (B) 6   (C) 6.8   (D) $4\sqrt{2}$ (E) $2\sqrt{3}$

Solution. Answer. (A)

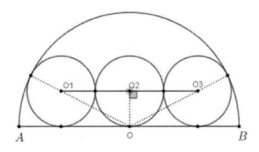

Let the centers of the three unit circles be O_1, O_2, and O_3. Let O be the center of the semicircle and R be its radius. Given that
$$OO_1 = OO_3 = R - 1.$$
It is not hard to show that the points O_1, O_2, and O_3 lie on the same line parallel to the line AB. Therefore, the line segment OO_2 is the median drawn from the vertex O to the base O_1O_3 of the isosceles triangle $\triangle OO_1O_3$. Thus, it follows that $OO_2 \perp AB$. Hence, using the Pythagorean Theorem for the right triangle $\triangle OO_1O_2$, we obtain that
$$(R-1)^2 = 1^2 + 2^2 = 5.$$
Thus, it follows that
$$R = 1 + \sqrt{5}.$$
Therefore
$$AB = 2R = 2 + 2\sqrt{5}.$$
□

Problem 4.47. *Let A be the set of natural numbers from 1 to 10 inclusive, and B be the set of natural numbers from 1 to 15 inclusive. Numbers a and b are randomly chosen from sets A and B, respectively. What is the probability that ab is divisible by 6?*

(A) $\dfrac{23}{50}$ (B) $\dfrac{1}{3}$ (C) $\dfrac{1}{5}$ (D) $\dfrac{11}{30}$ (E) $\dfrac{29}{150}$

Solution. Answer. (D)
The total number of pairs (a, b) is $10 \cdot 15 = 150$. The total number of favorable pairs (a, b) such that $6 \mid ab$ is
$$15 + 4 \cdot 5 + 2 \cdot 7 + 3 \cdot 2 = 55.$$
Thus, the required probability is equal to
$$\frac{55}{150} = \frac{11}{30}.$$
□

Problem 4.48. *The diagonals of convex quadrilateral $ABCD$ intersect at point O. The areas of $\triangle ABO$ and $\triangle CDO$ are 8 and 18, respectively. What is the smallest possible area of $ABCD$?*

(A) 39 (B) 50 (C) 52 (D) 40 (E) $10\sqrt{5}$

Solution. Answer. (B)

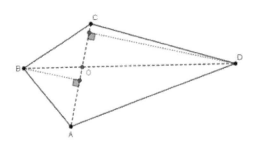

Note that the altitudes drawn from the vertex B in $\triangle ABO$ and $\triangle BOC$ are the same. Thus, it follows that
$$\frac{Area(ABO)}{Area(BOC)} = \frac{AO}{OC}.$$

Similarly, we obtain that
$$\frac{Area(ADO)}{Area(DOC)} = \frac{AO}{OC}.$$
Therefore
$$\frac{Area(ABO)}{Area(BOC)} = \frac{Area(ADO)}{Area(DOC)}.$$
Hence
$$Area(ADO) \cdot Area(BOC) = 8 \cdot 18 = 144.$$
We have that
$$Area(ABCD) = Area(ABO) + Area(BOC) + Area(CDO) + Area(ADO) =$$
$$= Area(ADO) + Area(BOC) + 26 = (\sqrt{Area(ADO)} - \sqrt{Area(BOC)})^2 + 50.$$
Thus, the least possible value for the area of the quadrilateral $ABCD$ is 50 (sq. units).
An example of such a quadrilateral (with area 50sq. units) is an isosceles trapezoid with bases AB and CD, diagonals of which are perpendicular to each other and
$$AO = BO = 4, CO = DO = 6.$$

$\square$

Problem 4.49. *In a regular 100-gon, a random diagonal is drawn. What is the probability that this diagonal is parallel to one of the polygon's sides?*

(A) $\dfrac{1}{97}$ (B) $\dfrac{1}{2}$ (C) $\dfrac{1}{3}$ (D) $\dfrac{2}{3}$ (E) $\dfrac{48}{97}$

Solution. Answer. (E)
The total number of diagonals in a regular 100-gon is
$$\frac{100 \cdot 97}{2} = 50 \cdot 97.$$
Note that number of diagonals parallel to a given side is
$$\frac{100 - 4}{2} = 48.$$
Thus, it follows that the required probability is equal to
$$\frac{50 \cdot 48}{50 \cdot 97} = \frac{48}{97}.$$

$\square$

Problem 4.50. *How many terms in the sequence $x_n = 10^n - 3^n + 2^n + 5$ are perfect squares?*

(A) 0   (B) 2   (C) 1   (D) 4   (E) 3

Solution. Answer. (C)
Note that $3 \mid 2^n + 1$ when n is odd and
$$3 \mid 10^n - 1,$$
for any n.

Therefore, for any odd n we have that
$$x_n = (10^n - 1) - 3^n + (2^n + 1) + 5$$
leaves a remainder 2 after a division by 3. Therefore, it cannot be a perfect square. We have that
$$x_2 = 10^2 - 3^2 + 2^2 + 5 = 10^2,$$
and for any positive integer k greater than 1 we have that
$$(10^k - 1)^2 < 10^{2k} - 3^{2k} + 2^{2k} + 5 < 10^{2k},$$
as
$$9^k - 4^k = (9-4)(9^{k-1} + \ldots + 4^{k-1}) > 5.$$

Therefore, x_{2k} is in between two consecutive perfect squares. Hence, it cannot be a perfect square. Thus, the only possibility for x_n to be a perfect square is when $n = 2$. $\square$

4.3 Solutions of AMC 10 type practice test 3

Problem 4.51. *There are 18 students in a 10th grade class. Five of them got 6 points on a recent test, four students got 7 points, six students got 8 points, and 3 students got 10 points. What is the average score for the test?*

(A) 7   (B) $7\frac{5}{9}$ (C) 6.5   (D) 8   (E) 9

Solution. Answer. (B)
The total score of the test is
$$5 \cdot 6 + 4 \cdot 7 + 6 \cdot 8 + 3 \cdot 10 = 136.$$
Therefore, the average score is
$$\frac{136}{18} = 7\frac{5}{9}.$$
□

Problem 4.52. *A square is divided into six squares (see the figure).*

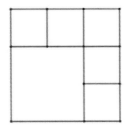

The side length of the largest of these six squares is 3. What is the side length of the original square?

(A) 9   (B) 5.5   (C) 5   (D) 4.5   (E) 6

Solution. Answer. (D)
The length of the smallest square is $3 : 2 = 1.5$. Therefore, the side length of the initial square is $3 \cdot 1.5 = 4.5$.
□

Problem 4.53. *Arthur planned to read a book in two days, and he was supposed to read three times as many pages on the first day as on the second day. However, he read 12 fewer pages than he originally planned. In order to finish reading the book in two days, Arthur read as many pages on the second day as on the first day. How many pages does the book have?*

(A) 40   (B) 48   (C) 64   (D) 16   (E) 32

Solution. Answer. (B)
Let the number of pages Arthur has planned to read on the first day and on the second day be $3x$ and x, respectively. Given that
$$3x - 12 = x + 12.$$
Therefore $x = 12$.
Thus, the total number of pages is $4x = 48$.
□

Problem 4.54. *Some trucks, each with a loading capacity of 1600 kilograms, have to transport 6.7 tonnes of sand. At least how many trucks are needed to transport the sand, if all the trucks should be loaded equally?*

(A) 4 (B) 6 (C) 5 (D) 7 (E) 8

Solution. Answer. (C)
4 tracks are not sufficient as 6.7(*tonnes*) : 4 = 1675 kilograms. In case of 5 tracks, each track could transport 1340 kilograms of sand. □

Problem 4.55. *Let two concentric circles form a ring with an area of 36π. What is the length of the longest line segment that can be drawn in this ring (between these two circles)?*

(A) 14 (B) 10 (C) 36 (D) 12 (E) 16

Solution. Answer. (D)

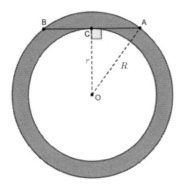

Note that a segment inscribed in the ring has the longest length, if it is tangent to the smaller circle. Therefore, if $OA = R$ and $OC = r$ are the radii of the large and the small circles, respectively. Then
$$\pi R^2 - \pi r^2 = 36\pi,$$
or equivalently
$$R^2 - r^2 = 36.$$
Thus, it follows that
$$AB = 2 \cdot AC = 2\sqrt{R^2 - r^2} = 2 \cdot 6 = 12.$$
□

Problem 4.56. *Let the operation $*$ be defined as follows: for any positive real numbers x and y, $x * y = xy - x - y$. What is x, if $x * x = 1 * (1 * 0)$?*

(A) $\sqrt{2}+1$ (B) 1 (C) $1-\sqrt{2}$ (D) 0 (E) $\sqrt{3}-\sqrt{2}$

Solution. Answer. (B)
Given that
$$x * x = x^2 - 2x,$$
$$1 * 0 = -1,$$
and
$$1 * (1 * 0) = 1 * (-1) = -1 - 1 - (-1) = -1.$$
Thus, it follows that
$$x^2 - 2x = -1.$$
Therefore $x = 1$. □

Problem 4.57. *In convex quadrilateral $ABCD$, $AB = 3$, $BC = CD = 1$, $\angle ABC = 150°$ and $\angle BCD = 120°$. What is the length of AD?*

(A) $\sqrt{3}$ (B) 2 (C) 1 (D) $\sqrt{3} + 1$ (E) 3

Solution. Answer. (E)

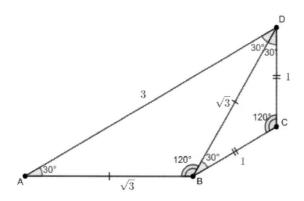

According to the law of cosines, we obtain that
$$BD^2 = BC^2 + CD^2 - 2 \cdot BC \cdot BD \cos 120°.$$

Thus, it follows that $BD = \sqrt{3}$. Note that
$$\angle CBD = \frac{180° - 120°}{2} = 30°,$$

and
$$\angle ABD = 150° - 30° = 120°.$$

According to the law of cosines again for $\triangle ABD$ we get
$$AD^2 = AB^2 + BD^2 - 2 \cdot AB \cdot BD \cos 120° = 9.$$

Thus, it follows that $AD = 3$. $\square$

Problem 4.58. *Two types of books, costing $15 and $26 were purchased for the school library. In all, 16 books were purchased. The total cost of the books was $339. How many books that cost $15 were purchased for the library?*

(A) 9 (B) 7 (C) 8 (D) 10 (E) 6

Solution. Answer. (B)
Let the quantity of $15 books be x. Then, the quantity of $26 books is $16 - x$. Therefore
$$15x + 26(16 - x) = 339.$$

Thus, it follows that
$$11x = 26 \cdot 16 - 339 = 77.$$

Hence, we obtain that $x = 7$. $\square$

Problem 4.59. *A palindrome number, such as 16061, is a number that remains the same when its digits are reversed. What is the smallest possible positive difference of two distinct four-digit palindromes?*

(A) 10   (B) 11   (C) 9   (D) 110   (E) 99

Solution. Answer. (B)
Let
$$\overline{abba} = 1001a + 110b = 11 \cdot (91a + 10b)$$
be a four-digit palindrome. Therefore, it is divisible by 11. Note that $2002 - 1991 = 11$. $\square$

Problem 4.60. *In a numeric sequence $x_1 = 1, x_2 = 5$ and $x_n = |x_{n-1} - x_{n-2}|, n = 3, 4, 5, ..., 2010$. What is the remainder when x_{2010} is divided by 7?*

(A) 1   (B) 2   (C) 0   (D) 4   (E) 3

Solution. Answer. (C)
Let us find some of the first terms:
$$x_3 = 4, x_4 = 1, x_5 = 3, x_6 = 2, x_7 = 1, x_8 = 1, x_9 = 0, x_{10} = 1, x_{11} = 1.$$

Thus, it follows that
$$x_{n+3} = x_n, n = 7, 8, ...$$

Hence, we deduce that
$$x_{2010} = x_{9+3 \cdot 667} = x_9 = 0.$$
$\square$

Problem 4.61. *The difference between positive numbers a and b is 4. The solution set to the inequality $b \leq 4 - x \leq a$ is a line segment with length 36. What is $a + b$?*

(A) 4   (B) 36   (C) 2   (D) 9   (E) 18

Solution. Answer. (D)
Let us solve the inequality
$$b \leq \sqrt{4-x} \leq a.$$

Thus, it follows that
$$b^2 \leq 4 - x \leq a^2.$$

Hence, we obtain that
$$4 - a^2 \leq x \leq 4 - b^2.$$

Given that
$$4 - b^2 - (4 - a^2) = 36.$$

Therefore
$$a^2 - b^2 = 36.$$

Thus, it follows that
$$a + b = \frac{a^2 - b^2}{a - b} = \frac{36}{4} = 9.$$
$\square$

Problem 4.62. *Given three spheres, the radii of the first two spheres are 3 and 4. The sum of the total surface area of these two spheres is equal to the total surface area of the third sphere. What is the volume of the third sphere?*

(A) 125π (B) 200π (C) 100π (D) $\dfrac{400\pi}{3}$ (E) $\dfrac{500\pi}{3}$

Solution. Answer. (E)
Let the radius of the third sphere be R. Given that
$$4\pi \cdot 3^2 + 4\pi \cdot 4^2 = 4\pi R^2.$$
Thus, it follows that $R = 5$. Hence, we deduce that the volume is equal to
$$\frac{4}{3}\pi \cdot R^3 = \frac{500\pi}{3}.$$
$\square$

Problem 4.63. *The distance between cities A and B is 320 miles. A car left City A and covered 120 miles during the first two hours of driving. After that, it stopped for half an hour and covered the rest of the trip driving at a speed of 80 mph. What was the car's average speed (in miles per hour) on the road between cities A and B?*

(A) 70   (B) 64   (C) 75   (D) 72   (E) 60

Solution. Answer. (B)
The car has spent $\dfrac{200}{80} = 2.5$ hours on the remaining trip. Therefore, the total time spent on the road AB is $2 + 0.5 + 2.5 = 5$ hours. Thus, the average speed is equal to $\dfrac{320}{5} = 64$ miles per hour. $\square$

Problem 4.64. *Let AC be the longest side of triangle ABC. Let BH be an altitude of triangle ABC. Given that $AC = 4BH$ and $\angle C = 15°$. What is the angle measure (in degrees) of $\angle A$?*

(A) 15   (B) 30   (C) 75   (D) 45   (E) 60

Solution. Answer. (C)

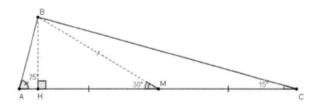

Let us choose the point M, such that
$$\angle MBC = \angle MCB = 15°.$$
Therefore $\angle HMB = 30°$. Thus, it follows that $\triangle HMB$ is a $30-60-90$ right triangle. Hence
$$MB = 2 \cdot BH = \frac{AC}{2}.$$
We also have that $MB = MC$. We obtain that
$$MA = AC - MC = AC - MB = \frac{AC}{2}.$$

105

Thus, it follows that
$$\angle A = \frac{180° - 30°}{2} = 75°.$$
□

Problem 4.65. *Each of the inhabitants of TruLi island $(A, B, C, D$ and $E)$ either always tells the truth or always lies. They once said the following:*
A said: "C is a truth-teller."
B said: "D is a truth-teller."
C said: "E is a truth-teller."
D said: "B is a truth-teller."
E said: "B is a liar."
What is the value of the product of the number of truth-tellers and the number of liars on TruLi island?

(A) 6 (B) 4 (C) 0 (D) 5 (E) 3

Solution. Answer. (A)
Let us consider the following two cases.
Case 1. B is a truth-teller, then as per B: D is a truth-teller, as per E: E is a liar, as per C: C is a liar, as per A: A is a liar. Therefore, there are 3 liars and 2 truth-tellers.
Case 2. B is a liar, then A, C, E are truth-tellers and B, D are liars.
Therefore, the answer is $3 \cdot 2 = 6$. □

Problem 4.66. *Let D be a point on side AC of triangle ABC. Given that $AB = 6, AC = 9, CD = 5$ and $BD = 8$. What is the value of the length of side BC?*

(A) $5\frac{1}{3}$ (B) 10 (C) 8 (D) 12 (E) 14

Solution. Answer. (D)

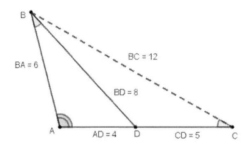

We have that
$$AD = AC - CD = 4.$$
Note that
$$\frac{AD}{AB} = \frac{AB}{AC}.$$
Therefore $\triangle ADB \sim \triangle ABC$. Thus, it follows that
$$\frac{BD}{BC} = \frac{4}{6}.$$
Hence, we obtain that $BC = 12$.
Alternative solution. As $AB = 6, AC = 9, CD = 5, BD = 8$ and $AD = 4$, then from Stewart's theorem, it follows that $BC = 12$. □

Problem 4.67. *Ben randomly chooses a number from each of the sets: $\{1, 3, 5, 7, 9, 11\}$ and $\{2, 4, 6, 8, 10\}$. What is the probability that the sum of the chosen numbers is a multiple of 3?*

(A) $\dfrac{4}{15}$ (B) $\dfrac{1}{3}$ (C) $\dfrac{1}{5}$ (D) $\dfrac{1}{4}$ (E) $\dfrac{1}{6}$

Solution. Answer. (B)

Suppose Ben did choose numbers a and b from the 1st (first six odd numbers) and 2nd (first five even numbers) sets, respectively. Therefore, the total number of all possible (a, b) pairs is equal to $6 \cdot 5 = 30$, while the total number of all favorable outcomes is 10 as it can be seen from the "+" table below:

+	1	3	5	7	9	11
2	3	5	7	9	11	13
4	5	5	9	11	13	15
6	7	9	11	13	15	17
8	9	11	13	15	17	19
10	11	13	15	17	19	21

Therefore, the required probability is $\dfrac{10}{30} = \dfrac{1}{3}$. □

Problem 4.68. *Let the edge length of a cube be 4 inches. A right circular cylinder and a square prism are removed from this cube (see the figure). Given that the radius of the base of the cylinder is 1 inch and the centers of its bases coincide with the centers of two opposite faces of the cube. The centers of the bases of the square prism coincide with the centers of two other opposite faces of the cube. The base edge length of the square prism is 2 inches and all its lateral faces are parallel to the corresponding faces of the cube. What is the value of the volume (in cubic inches) of the remaining solid?*

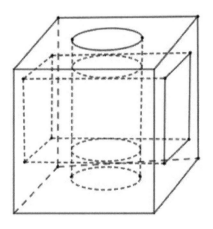

(A) $56 - 4\pi$ (B) $64 - 4\pi$ (C) 48 (D) 32 (E) $48 - 2\pi$

Solution. Answer. (E)

Note that the value of the volume (in cubic inches) of the remaining solid is:

$$4^3 - 2 \cdot 2 \cdot 4 - 2 \cdot \pi \cdot 1^2 \cdot 1 = 48 - 2\pi.$$

□

Problem 4.69. *Let $ABCDEF$ be a convex hexagon, such that $\angle B = \angle D = \angle F = 120°$, $\angle C = 2 \cdot \angle ACE$ and $\angle A = 2 \cdot \angle CAE$. Given that the area of triangle $ACE = 10\sqrt{3}$ and the area of hexagon $ABCDEF = 11\sqrt{3}$. What is the value of $|DE - EF|$?*

(A) $\sqrt{3}$ (B) 3   (C) $3\sqrt{3}$ (D) 6   (E) $6\sqrt{3}$

Solution. Answer. (D)

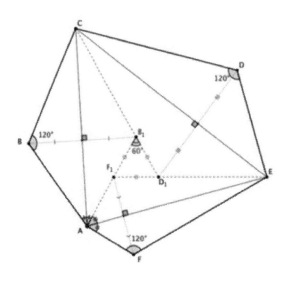

Note that
$$\angle E = 360° - \angle A - \angle C = 360° - 2\angle ACE - 2\angle CAE = 2\angle AEC.$$

Thus, it follows that $\angle E = 2\angle AEC$.

Let us consider $\triangle ABC, \triangle CDE,$ and $\triangle AEF$ congruent triangles with respect to lines $AC, CE,$ and AE (as it is shown in the figure). Note that $\triangle B_1 D_1 F_1$ is an equilateral triangle with area
$$10\sqrt{3} - (11\sqrt{3} - 10\sqrt{3}) = 9\sqrt{3}.$$

On the other hand, the same area is equal to
$$\frac{\sqrt{3}}{4} F_1 D_1^2.$$

We obtain that
$$F_1 D_1^2 = 36.$$

Therefore
$$F_1 D_1 = 6.$$

Thus, it follows that
$$|DE - EF| = |D_1 E - EF_1| = F_1 D_1 = 6.$$

$\square$

Problem 4.70. *A fly is sitting on one of the vertices of a unit cube. What is the shortest possible length of the path it can fly, if it needs to visit each vertex of the cube?*

(A) 8   (B) 6   (C) 7   (D) $6 + \sqrt{3}$ (E) $5 + \sqrt{3}$

Solution. Answer. (C)
Assume that initially the snail is at vertex A and the first vertex it visits is A_1. Then, the next vertices it visits are A_2, A_3 and so on until the last one A_7. Obviously, the total length of the path is at least

$$A_1A_2 + A_2A_3 + A_3A_4 + A_4A_5 + A_5A_6 + A_6A_7 \geq 1 + 1 + \ldots + 1 = 7.$$

Below is provided an example of the path of length 7.

□

Problem 4.71. *Given nonzero numbers a, b, and c, the polynomials $x^3 + ax^2 + bx + c, x^2 + bx + c$, and $x^2 + ax + c$ have a common root. What is the value of $\dfrac{20a + 10b}{a + b}$?*

(A) 15   (B) 10   (C) 20   (D) 14   (E) 12

Solution. Answer. (A)
Let x_0 be the common root of the polynomials. Then

$$x_0^3 + ax_0^2 + bx_0 + c = 0,$$

$$x_0^2 + bx_0 + c = 0,$$

and

$$x_0^2 + ax_0 + c = 0.$$

As

$$x_0^2 + bx_0 + c = 0,$$

then

$$bx_0 + c = -x_0^2,$$

or

$$x_0^3 + ax_0^2 - x_0^2 = 0.$$

So, $x_0 = 0$ or $x_0 = 1 - a$. Note that $x_0 \neq 0$, as in that case $c = 0$, which is not possible. Hence, $x_0 = 1 - a$. Therefore

$$(1 - a)^2 + a(1 - a) + c = 0,$$

or $c = a - 1$. As well as

$$(1 - a)^2 + b(1 - a) + a - 1 = 0.$$

Therefore $1 - a + b - 1 = 0$ or $a = b$.
Thus, the answer is

$$\frac{20a + 10b}{a + b} = 15.$$

For example, when $a = b = 2, c = 1$, we have that $x_0 = -1$.   □

Problem 4.72. *Given a regular octagon, how many isosceles triangles whose vertices coincide with the vertices of the octagon exist?*

(A) 16   (B) 48   (C) 64   (D) 32   (E) 24

Solution. Answer. (E)

Let us first construct a circumcircle to the octagon and note that there is no equilateral triangle with vertices being the vertices of the octagon. As well as, each vertex of the octagon is an opposite vertex to the base of exactly three isosceles triangles. Therefore, there are $8 \cdot 3 = 24$ such triangles. $\square$

Problem 4.73. *Sam is randomly choosing n unit squares from an $n \times n$ square. The probability that the n chosen unit squares are in different columns and in different rows is $\dfrac{5}{324}$. What is n ?*

(A) 39   (B) 50   (C) 52   (D) 40   (E) $10\sqrt{5}$

Solution. Answer. (D)
According to the Multiplication Rule, there are total n^n possibilities. As well as, there are $n(n-1)(n-2)...1 = n!$ number of favorable outcomes. Therefore
$$\frac{n!}{n^n} = \frac{(n-1)!}{n^{n-1}} = \frac{5}{324}.$$
Thus, if $n = 6$, then
$$\frac{(n-1)!}{n^{n-1}} = \frac{5!}{6^5} = \frac{5}{324}.$$
If $n \geq 7$, then
$$\frac{(n-1)!}{n^{n-1}} = \frac{5!}{n^5} \cdot \frac{6}{n} \cdot .... \cdot \frac{n-1}{n} < \frac{5!}{n^5} \cdot 1 \cdot .... \cdot 1 < \frac{5!}{6^5} = \frac{5}{324}.$$
Easy check shows that if $n \leq 5$, then
$$\frac{(n-1)!}{n^{n-1}} > \frac{5!}{6^5}.$$

$\square$

Problem 4.74. *What is the value of the sum of the last two digits of the sum $1^3 + 2^3 + ... + 2010^3$?*

(A) 10   (B) 7   (C) 9   (D) 5   (E) 17

Solution. Answer. (B)
Using the identity
$$a^3 + b^3 = (a+b)(a^2 - ab + b^2),$$
we obtain that
$$100 \mid (-10)^3 + 2010^3, ..., 100 \mid 0^3 + 2000^3, 100 \mid 1^3 + 1999^3, 100 \mid 2^3 + 1998^3, ..., 100 \mid 999^3 + 1001^3.$$
So, the last two digits of the required sum coincide with that of
$$1^3 + 2^3 + ... + 10^3 + 1000^3.$$
The last two digits of which end in 25. Therefore, the sum of the digits is $2 + 5 = 7$. □

Problem 4.75. *What is the smallest possible positive integer value of n, such that $0 < \{\sqrt[3]{n}\} < \frac{1}{99}$? Here, $\{x\}$ denotes the fractional part of a real number x.*

(A) 196   (B) 210   (C) 217   (D) 187   (E) 143

Solution. Answer. (C)
Let k be a positive integer, such that
$$k^3 < n < (k+1)^3.$$
Given that
$$\sqrt[3]{n} - [\sqrt[3]{n}] < \frac{1}{99}.$$
Thus, it follows that
$$\sqrt[3]{n} < k + \frac{1}{99}.$$
Therefore
$$n < k^3 + \frac{k^2}{33} + \frac{3k}{99^2} + \frac{1}{99^3},$$
as well as $n \geq k^3 + 1$. Hence
$$\frac{k^2}{33} + \frac{3k}{99^2} + \frac{1}{99^3} \geq 1,$$
or $k \geq 6$. Thus, it follows that $n \geq 6^3 + 1 = 217$. On the other hand, we have that
$$\{\sqrt[3]{217}\} = \sqrt[3]{217} - [\sqrt[3]{217}] = \sqrt[3]{217} - 6 = \frac{1}{\sqrt[3]{217}^2 + 6\sqrt[3]{217} + 6^2} < \frac{1}{6^2 + 6^2 + 6^2} < \frac{1}{99}.$$
Hence, we obtain that
$$0 < \{\sqrt[3]{217}\} < \frac{1}{99}.$$
□

4.4 Solutions of AMC 10 type practice test 4

Problem 4.76. *A pen costs $0.5. When buying 10 pens you receive a discount of $1. How much will 10 pens cost?*

(A) $5   (B) $4   (C) $4.5   (D) $3   (E) $3.5

Solution. Answer. (B)
We have that $10 \cdot 0.5 - 1 = \$4$. □

Problem 4.77. *The cost of renting a club at a driving range is $1.5. If the owner charges $0.25 for each used ball, what is the greatest number of shots that one can take for $10?*

(A) 34   (B) 35   (C) 40   (D) 36   (E) 32

Solution. Answer. (A)
We have that $(10 - 1.5) : 0.25 = 34$. □

Problem 4.78. *ABCD is a square with a side length of 6. Point E lies on the line segment $\overline{BC}$. What is the area of $\triangle AED$?*

(A) 30   (B) 24   (C) 12   (D) 18   (E) 6

Solution. Answer. (D)
Note that perpendicular drawn from point E on side AD has a length of 6. Thus, it follows that

$$Area(\triangle AED) = \frac{1}{2} \cdot 6 \cdot 6 = 18.$$

□

Problem 4.79. *Among ten children on a hiking tour, any two of them have different quantities of candy. They split up equally into two groups, and it turns out that the total amount of candy in the first group is five times smaller than the total amount of candy in the second group. What is the smallest possible total amount of candy the children can have?*

(A) 70   (B) 50   (C) 60   (D) 40   (E) 30

Solution. Answer. (C)
Note that total quantity of candies in the 1st group is greater than or equal to $0+1+2+3+4 = 10$ and the total quantity of candies all the children have is greater than or equal to $10 + 10 \cdot 5 = 60$. A possible "scenario" of the total number of candies 60 and such that in the 1st group the number of candies is 10 could be $\{0, 1, 2, 3, 4\}$ and $\{7, 8, 9, 10, 16\}$. □

Problem 4.80. *There are 112 apples in one box, 97 apples in a second box, and 88 apples in a third box. First, a apples from the first box were transferred to the second box, and then b apples were transferred from the second box to the third box. After that, all the boxes had an equal number of apples. What is a + b?*

(A) 11   (B) 13   (C) 20   (D) 25   (E) 24

Solution. Answer. (E)
Note that the quantity of apples in all the three boxes is equal (after the last transfer of the apples from the 2nd box to the 3rd box). The quantity of apples in each box was

$$\frac{112 + 97 + 88}{3} = 99.$$

Thus, it follows that
$$a = 112 - 99 = 13,$$
and
$$b = 97 + 13 - 99 = 11.$$
Therefore
$$a + b = 13 + 11 = 24.$$
□

Problem 4.81. *Aram's four children are now 1, 5, 7, and 9 years old. In how many years from now will the sum of the ages of two of his children be twice the sum of the ages of the other two children?*

(A) 1   (B) 2   (C) 3   (D) 5   (E) 10

Solution. Answer. (B)
In x years from now, the ages of children will be $1+x, 5+x, 7+x$, and $9+x$. Given that $7+x+9+x = 2(1+x+5+x)$. Therefore $x = 2$. □

Problem 4.82. *There are six seats in a boat. In how many ways can three girls and five boys be seated in the boat so that there are at least two girls?*

(A) 15   (B) 25   (C) 10   (D) 20   (E) 18

Solution. Answer. (B)
There are the following possible seating arrangements:
1) 2 girls and 4 boys or
2) 3 girls and 3 boys.
Therefore, the total number of possible arrangements is equal to
$$\binom{3}{2} \cdot \binom{5}{4} + \binom{3}{3} \cdot \binom{5}{3} = 25.$$
□

Problem 4.83. *What is x, if $2^{2013} : 2^x = 2^{25} \cdot 2^{1975}$?*

(A) 0   (B) $\dfrac{2013}{2010}$ (C) 4013   (D) 13   (E) -13

Solution. Answer. (D)
We have that
$$2^{2013-x} = 2^{25+1975}.$$
Thus, it follows that $2013 - x = 2000$. Therefore $x = 13$. □

Problem 4.84. *There are 22 students in a class. 20% of the boys and 25% of the girls in the class are "A" grade students. How many "A" grade students are there in the class?*

(A) 4   (B) 2   (C) 5   (D) 3   (E) 6

Solution. Answer. (C)
Let us denote by m the number of boys in the class and by n the number of girls in the class. Thus, it follows that $m + n = 22$. We have that $\dfrac{m}{5}$ boys and $\dfrac{n}{4}$ girls are A grade students. Therefore $5 \mid m$ and $4 \mid n$. Hence, we obtain that $m = 10$ and $n = 12$. Thus, it follows that $\dfrac{m}{5} + \dfrac{n}{4} = 5$. □

Problem 4.85. *Mary read a book in four days. The number of pages she read on the second day was two times less than the number of pages she read on the first day, while the number of pages she read on the fourth day was 50% of the number of pages she read on the second day. The ratio of the number of pages she read on the third day to the number of pages she read on the fourth day is 3 : 4. What percent of the book did Mary read on the first day?*

(A) $\dfrac{20}{3}$ (B) 40 (C) 70 (D) $\dfrac{1600}{31}$ (E) 30.3

Solution. Answer. (D)

Let the number of pages Mary read on the 1st day be a pages. Therefore, on the 2nd day she read $\dfrac{a}{2}$ pages, on the 4th day $\dfrac{a}{4}$ pages, and on the 3rd day

$$\frac{3}{4} \cdot \frac{a}{4} = \frac{3a}{16}$$

pages. Thus, it follows that on the 1st day, she read

$$\frac{a}{a + \dfrac{a}{2} + \dfrac{a}{4} + \dfrac{3a}{16}} \cdot 100\% = \frac{1600}{31}\%.$$

□

Problem 4.86. *Monday's schedule of a 10th grade class should consist of several distinct subjects. The total number of possibilities to distribute any two of the subjects for the first two classes on Monday is 20. In how many different ways can Monday's schedule be made using all the subjects?*

(A) 50 (B) 720 (C) 24 (D) 60 (E) 120

Solution. Answer. (E)

Let the number of the subjects for Monday's schedule be x. Therefore, the total number of ways to distribute the first two classes is $x(x-1) = 20$. Thus, it follows that $x = 5$. Hence, in total there are $5! = 120$ ways to make Monday's schedule. □

Problem 4.87. *A straight line, containing incenter I of $\triangle ABC$ and parallel to side AC, intersects sides AB and BC at points M and N, respectively. What is the perimeter of trapezoid AMNC, if MN=10 and AC=15?*

(A) 35 (B) 40 (C) 30 (D) 50 (E) 55

Solution. Answer. (A)

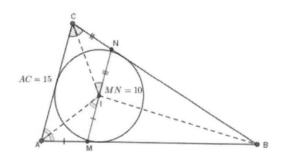

As I is the incenter of $\triangle ABC$, then $\angle IAB = \angle IAC$ and $MN \parallel AC$. Thus, it follows that

$$\angle MAI = \angle IAB = \angle IAC = \angle AIM.$$

Therefore $AM = MI$. Similarly, we obtain that $CN = NI$. Hence, we deduce that the perimeter of $AMNC$ is
$$AM + MN + NC + AC = MI + MN + NI + AC = 2MN + AC = 35.$$
□

Problem 4.88. *How many three-digit multiples of 11 exist such that the sum of the digits is less than 11?*

(A) 16   (B) 15   (C) 10   (D) 14   (E) 12

Solution. Answer. (B)
Let
$$\overline{abc} = 100a + 10b + c$$
be a three-digit number, such that
$$11 \mid \overline{abc},$$
and
$$a + b + c < 11.$$
Hence, from
$$11 \mid 99a + 11b + a - b + c$$
it follows that
$$11 \mid a - b + c,$$
and
$$-10 < a - b + c < 11.$$
Therefore, we obtain that
$$a - b + c = 0.$$
Thus, we obtain that
$$a + c = b, b \in \{1, 2, 3, 4, 5\}.$$
Hence, in total there are $1 + 2 + 3 + 4 + 5 = 15$ numbers.   □

Problem 4.89. *The sum of the number of sides, faces, and vertices of a prism is between 2012 and 2024. How many faces does the prism have?*

(A) 338   (B) 2016   (C) 337   (D) 339   (E) 350

Solution. Answer. (A)
Suppose the prism is n-gon-prism. Therefore, the number of sides, faces, and vertices are $3n, n+2$, and $2n$, respectively. Given that
$$2012 < 6n + 2 < 2024.$$
Thus, it follows that $335 < n < 337$. Hence, we obtain that $n = 336$. Therefore $n + 2 = 338$.   □

Problem 4.90. *The lengths of the altitudes of a triangle are three consecutive terms of a geometric sequence. What is the length of the third side of the triangle, if it is given that the other two sides have lengths of 4 and 9 units?*

(A) 7   (B) 6.5   (C) 6   (D) 7.5   (E) 8

Solution. Answer. (C)
Let h_1, h_2, and h_3 be the altitudes corresponding to the sides a, b, and c. The area of the triangle is $\frac{ah_1}{2} = \frac{bh_2}{2} = \frac{ch_3}{2}$, we obtain that $a : b = h_2 : h_1 = h_3 : h_2 = b : c$. Therefore, if h_1, h_2, and h_3 form a geometric sequence, then a, b, c form a geometric series as well. Thus, it follows that $\{a, b, c\} \in \{\{4, 6, 9\}, \{\frac{16}{9}, 4, 9\}, \{4, 9, \frac{81}{4}\}\}$. Among these triples, only $4, 6, 9$ satisfies the triangle inequality. $\square$

Problem 4.91. *Points $A(1,1)$ and $C(7,4)$ are reflected over the line $y = 2$. What is the area of the quadrilateral whose vertices are the two given points and the two obtained points?*

(A) 10.5   (B) 13   (C) 20   (D) 17   (E) 18

Solution. Answer. (E)

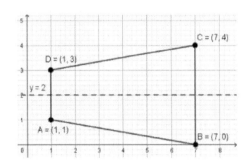

Note that $AD \parallel BC$ and $D(1,3), B(7,0)$. Therefore, the quadrilateral $ABCD$ is a trapezoid with bases 2 and 4 and altitude 6. Thus, it follows that

$$Area(ABCD) = \frac{2+4}{2} \cdot 6 = 18.$$

$\square$

Problem 4.92. *Given the following three arithmetic sequences : 1, 3, ... and 1, 4, ... and 1, 5, ... A number is randomly selected from 1 to 1000, inclusive. What is the probability that the selected number is not a term of any of the given sequences?*

(A) 0.33   (B) 0.3   (C) $\frac{101}{1000}$ (D) 0.5   (E) 0.333

Solution. Answer. (E)
First note that numbers in the form $6k$, where $k \geq 1$ and $6k+2$, where $k \geq 0$, such that they are between 1 and 1000 are not the terms of given arithmetic sequences. There are respectively

$$\lfloor \frac{1000}{6} \rfloor = 166,$$

and $166 + 1 = 167$ numbers. Therefore, in total there are 333 numbers having that property. $\square$

Problem 4.93. *Convex quadrilateral $ABCD$, with an area of 18 and vertex coordinates $A(1,1), B(2,7), C(m,n), D(6,3)$ is drawn on the xy–coordinate plane. What is $m+n$?*

(A) 12   (B) 7   (C) 8   (D) 11   (E) 10

Solution. Answer. (D)

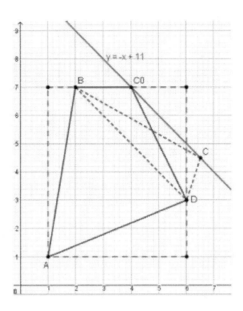

Let us consider the point $C_0(4,7)$. Note that

$$Area(ABC_0D) = 5 \cdot 6 - \frac{6 \cdot 1}{2} - \frac{5 \cdot 2}{2} - 4 \cdot 22 = 18 = Area(ABCD).$$

Therefore, $CC_0 \parallel BD$ and the equation of line BD is $y = -x + 9$. Taking this into consideration, and as line CC_0 is line BD parallelly moved by 2 (see the figure), then the equation of line CC_0 is $y = -x + 11$ or $m + n = 11$. □

Problem 4.94. *How many natural numbers n have the following property: 2013 leaves a remainder of 13 when divided by n^2?*

(A) 6   (B) 4   (C) 10   (D) 8   (E) 5

Solution. Answer. (B)
Given that $n^2 > 13$ and $2013 - 13 = 2000$ is divisible by n^2. On the other hand

$$2000 = 2^4 \cdot 5^3.$$

Thus, it follows that $n \mid 5 \cdot 4$. Therefore

$$n \in \{4, 5, 10, 20\}.$$

Hence, four natural numbers n have the given property. □

Problem 4.95. *A square with a side length of 2 is rotated around one of its vertices by 30°. What is the area of the part of the plane formed by the square and its rotation?*

(A) $4 + \dfrac{2\pi}{3}$ (B) 4   (C) 6   (D) 2π (E) $\pi + 3$

Solution. Answer. (A)

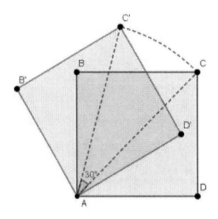

The area of the figure is
$$2^2 + \frac{\pi(2\sqrt{2})^2}{360} \cdot 30 = 4 + \frac{2\pi}{3}.$$

□

Problem 4.96. *A painter painted $\dfrac{1}{3}$ of a wall on the first day of a job. On the second day, he painted $\dfrac{1}{5}$ of the remaining part of the wall, and so on. On the nth day he painted $\dfrac{1}{2n+1}$ of the part of the wall that remained after the $(n-1)^{th}$ day. What part of the wall still needed to be painted after the 6th day?*

(A) $\dfrac{1}{3}$ (B) $\dfrac{1}{125}$ (C) $\dfrac{512}{3003}$ (D) $\dfrac{1024}{3003}$ (E) $\dfrac{2048}{3003}$

Solution. Answer. (D)

1st day the painter has painted $\dfrac{1}{3}$ of the wall and $\dfrac{2}{3}$ has remained.

2nd day the painter has painted $\dfrac{1}{5} \cdot \dfrac{2}{3}$ and $\dfrac{2}{3} \cdot \dfrac{4}{5}$ has remained.

and so on, 6th day has remained
$$\frac{2}{3} \cdot \frac{4}{5} \cdot \frac{6}{7} \cdot \frac{8}{9} \cdot \frac{10}{11} \cdot \frac{12}{13} = \frac{1024}{3003}.$$

□

Problem 4.97. *The centers of eight congruent spheres with radii 1 are vertices of a cube with a side length of 3. What is the radius of a sphere that is inside the cube and tangent to the eight given spheres?*

(A) $1.5\sqrt{3}$ (B) 1 (C) $1.5\sqrt{3} - 1$ (D) $1.5\sqrt{3} + 1$ (E) 3

Solution. Answer. (C)

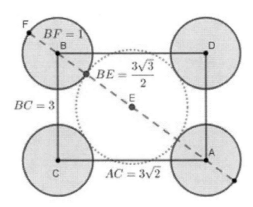

Let us denote the radius of the circumscribing sphere that is tangent to given eight spheres by r and its center by E. Note that the distance from E to each of the vertices of the cube is the same. Therefore, E is the center of the sphere circumscribing the cube. Thus, it follows that

$$EB = \frac{3\sqrt{3}}{2}.$$

Hence, we obtain that

$$r = 1.5\sqrt{3} - 1.$$

□

Problem 4.98. *A circle passing through vertices A and C of triangle ABC, also intersects sides AB and BC at points M and N, respectively. The side lengths of triangle BMN are integers and $AB = 8, BC = 6$. What is the largest possible length of line segment MN?*

(A) 10 (B) 9 (C) 8 (D) 7 (E) 6

Solution. Answer. (E)

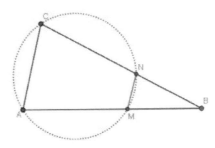

Using the Power of a Point Theorem, we obtain that

$$BM \cdot AB = BN \cdot BC.$$

Therefore $3BN = 4BM$. Hence $BN = 4, BM = 3$.

Taking this into consideration, and from the Triangle inequality $(BM + BN > MN)$ it follows that the largest possible length of line segment MN is 6. Note that for the triangle ABC with sides

$$AB = 8, BC = 6, AC = 12,$$

if

$$BM = 3, BN = 4,$$

then the points A, M, N, and C lie on the same circle. From the similarity of the triangles ABC and BMN, we obtain that $MN = 6$.

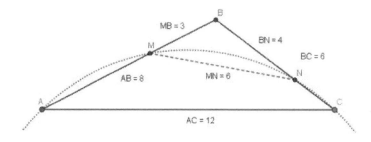

□

Problem 4.99. *The digits of a five-digit number are selected randomly, and the second digit from the right is even. What is the probability that the resulting five-digit number is divisible by 12?*

(A) 0.5   (B) 0.1   (C) 0.4   (D) $\dfrac{2}{3}$ (E) $\dfrac{1}{3}$

Solution. Answer. (B)

Note that, if we choose the first 4 digits of the five-digit number, then the last digit is possible to choose only in one way. Indeed, if the rightmost digit is even and is divisible by 4, then it can be either 0, 4 or 8. If the sum of the first four digits is S, then among the numbers $S, S+4, S+8$ only one is divisible by 3. Therefore, the rightmost digit is defined uniquely. Thus, the required probability is

$$\frac{9 \cdot 10 \cdot 10 \cdot 5 \cdot 1}{9 \cdot 10 \cdot 10 \cdot 5 \cdot 10} = 0.1.$$

□

Problem 4.100. *At least in how many points, different from the vertices of the hexagon, the diagonals of the convex hexagon can intersect?*

(A) 13   (B) 14   (C) 12   (D) 15   (E) 11

Solution. Answer. (A)

Note that the diagonals of the regular hexagon intersect at 13 points. Let us prove that the number of intersections of the diagonals for any convex hexagon is not less than 13. Indeed, for every 4 vertices there is 1 intersection point of diagonals. Moreover, there exists only one triple of diagonals, such that they intersect at 1 point. Hence, the number of intersections of the diagonals is not less than

$$\binom{6}{2} - \left(\binom{3}{2} - 1\right) = 15 - 2 = 13.$$

□

4.5 Solutions of AMC 10 type practice test 5

Problem 4.101. *What is the value of $(3^{-1} + 6^{-1} - 4^{-1})^{-1} : 5^0$?*

(A) $\dfrac{1}{4}$ (B) 4   (C) 5   (D) 1   (E) $\dfrac{4}{5}$

Solution. Answer. (B)
We have that
$$(3^{-1} + 6^{-1} - 4^{-1})^{-1} : 5^0 = \Big(\dfrac{1}{3} + \dfrac{1}{6} - \dfrac{1}{4}\Big)^{-1} : 1 = 4.$$
□

Problem 4.102. *There are triangles and pentagons drawn on the plane, the total number of which is 11. The total sum of all the internal angles of these 11 figures is 3420°. How many triangles are drawn on the plane?*

(A) 11   (B) 0   (C) 5   (D) 7   (E) 4

Solution. Answer. (D)
Let us split each of the pentagons (by one of the diagonals) to a triangle and a quadrilateral. The sum of all internal angles of 11 triangles is $11 \cdot 180° = 1980°$. Therefore, the sum of all internal angles of all the quadrilaterals is $3420° - 1980° = 1440°$. Thus, it follows that the total number of quadrilaterals is $1440° : 360° = 4$. Hence, the total number of triangles is $11 - 4 = 7$.   □

Problem 4.103. *Fig. 1 is made of unit length sticks. Fig. 2 is made from Fig. 1 by removing 8 equilateral triangles, with side lengths of one unit.*

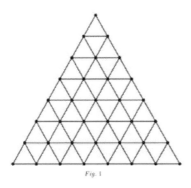

How many sticks are there in Fig. 2?

(A) 72   (B) 82   (C) 123   (D) 102   (E) 51

Solution. Answer. (A)
First note that the total number of equilateral triangles with a unit side is
$$(1 + 2 + ... + 7) + (1 + 2 + ... + 6) = 49.$$

In Fig. 2, we have $7 \cdot 3 = 21$ sticks are sides of separate triangles, while the other sticks are sides of exactly two triangles. Therefore, the total number of sticks in Fig. 2 is
$$\dfrac{41 \cdot 3 - 21}{2} + 21 = 72.$$
□

Problem 4.104. *A father noticed that the sum of the ages of his three children is greater by 8 than the sum of their ages 3 years ago. What is the age of his youngest child, if the children's ages are whole numbers?*

(A) 1   (B) 3   (C) 4   (D) 5   (E) 2

Solution. Answer. (E)
Two of his children should have grown up by 3 years and one by 2 years, so the youngest child is 2 years old. □

Problem 4.105. *There are 7 boys and 8 girls in a class. The teacher gave each of these 15 students a test with 10 problems. Each problem is scored 1 or 0 points. It is known that the average score of the girls was 7, while the average score of the class was 5.6. What was the average score of the boys?*

(A) 4.2   (B) 5   (C) 4   (D) 6   (E) 8

Solution. Answer. (C)
The total score of the class was $15 \cdot 5.6 = 84$, while the total score of the girls was $8 \cdot 7 = 56$. Therefore, the total score of the boys was $84 - 56 = 28$. Thus, it follows that the average score of the boys was $\frac{28}{7} = 4$. □

Problem 4.106. *The ratio of the sum of the cubes of two positive numbers to the difference of the cubes of these numbers is $\frac{189}{61}$. By what percent is the larger number greater than the smaller number?*

(A) 50   (B) 25   (C) 100   (D) 40   (E) 60

Solution. Answer. (B)
Let x and y be considered positive numbers. Given that

$$\frac{x^3 + y^3}{x^3 - y^3} = \frac{189}{61}.$$

Thus, it follows that
$$128x^3 = 250y^3.$$

Therefore
$$x = \frac{5}{4}y.$$

Hence, we obtain that
$$\frac{\frac{5}{4}y - y}{y} \cdot 100\% = 25\%.$$

□

Problem 4.107. *How many two-digit numbers leave a remainder of 2 when divided by 5?*

(A) 18   (B) 19   (C) 17   (D) 90   (E) 20

Solution. Answer. (A)
Note that the first two-digit number that leaves remainder 2 when divided by 5 is 12, then $12 + 5 = 17$, and $17 + 5 = 22$, etc. until the last one: 97. If we subtract 7 from each of these numbers and divide the result by 5, we get 1, 2, 3, ..., 18. Therefore, 18 numbers. □

Problem 4.108. *Five years ago Jane was three times Mia's age, while the sum of their ages now is 34. How old is Mia?*

(A) 6 (B) 10 (C) 13 (D) 11 (E) 8

Solution. Answer. (D)
Let now the age of Jane be x and the age of Mia be y. Given that
$$x + y = 34,$$
and
$$x - 5 = 3(y - 5).$$
Thus, it follows that
$$3(y - 5) + 5 + y = 34.$$
Therefore $y = 11$. Hence, Mia is 11 years old. $\square$

Problem 4.109. *Points A and B lie in the first quadrant of the coordinate plane and belong to the graph of $y = \dfrac{1}{x^3}$. The abscissa of point A is 25% bigger than the abscissa of point B. By what percent is the ordinate of point A smaller than the ordinate of point B?*

(A) 75 (B) 25 (C) 20 (D) 50 (E) 48.8

Solution. Answer. (E)
Let x_0 be the abscissa of the point B. Thus, the abscissa of the point A is $\dfrac{5}{4}x_0$. The ordinates of the points A and B are
$$\frac{64}{125x_0^3},$$
and
$$\frac{1}{x_0^3}.$$
Therefore
$$\frac{1 - \dfrac{64}{125}}{1} \cdot 100\% = 48.8\%.$$
$\square$

Problem 4.110. *How many three-digit numbers have the following property: the non-negative difference of any two neighboring digits is not less than 8?*

(A) 8 (B) 6 (C) 7 (D) 9 (E) 10

Solution. Answer. (C)
Let $\overline{abc}$ be a number with the property described in the problem, therefore $|a - b| \geq 8$ and $|b - c| \geq 8$. Thus, it follows that $a \in \{1, 8, 9\}$.
If $a = 1$, then $b = 9$ and either $c = 0$ or $c = 1$.
If $a = 8$, then $b = 0$ and either $c = 8$ or $c = 9$.
If $a = 9$, then either $b = 0$ or $b = 1$. In the case when $b = 0$, we have that either $c = 8$ or $c = 9$. In the case when $b = 1$, we have that $c = 9$.
Therefore, in total there are $3 \cdot 2 + 1 = 7$ such numbers. $\square$

Problem 4.111. *For positive numbers x and y, $x^2 - 2xy - 3y^2 = 0$. What is the value of $\dfrac{x^2 + 3y^2}{xy}$?*

(A) 1 (B) 4 (C) 2 (D) 5 (E) 1.5

Solution. Answer. (B)
We have that
$$x^2 - 2xy - 3y^2 = x^2 - y^2 - 2xy - 2y^2 = (x-y)(x+y) - 2y(x+y) = (x-3y)(x+y),$$
and $x + y > 0$. Thus, it follows that $x - 3y = 0$, if
$$x^2 - 2xy - 3y^2 = 0.$$
Therefore $x = 3y$. Hence, we obtain that
$$\frac{x^2 + 3y^2}{xy} = \frac{12y^2}{3y^2} = 4.$$
□

Problem 4.112. *Let a, b, and c be positive real numbers such that the point (a,b) belongs to the graph of $y = x^2 + 2x + 2$ and the point $(c, b+3)$ belongs to the graph of $y = x^2 - 2x + 5$. What is $c-a$?*

(A) 1 (B) -2 (C) 2 (D) 0 (E) 3

Solution. Answer. (C)
Given that
$$b = a^2 + 2a + 2,$$
and
$$b + 3 = c^2 - 2c + 5.$$
Thus, it follows that
$$c^2 - 2c = a^2 + 2a.$$
Therefore
$$(c - a - 2)(c + a) = 0.$$
We have that $c + a > 0$, thus $c - a = 2$. □

Problem 4.113. *How many integers from 1 to 100 can be represented in the form $3m + 5n$, where m and n are whole numbers such that $m + n \leq 20$?*

(A) 94 (B) 96 (C) 97 (D) 95 (E) 50

Solution. Answer. (A)
We have that $20 \cdot 5 = 100$. Note that if the number of 5's is increased by 1 and the number of 3's increased by 1, then the value of $3m + 5n$ is decreased by 2. Therefore, numbers 60, 62, 64, ..., 100 are of this form. We have that $19 \cdot 5 = 95$. Therefore, 57, 59, ..., 95 also have this form. Similarly 3, 6, ..., $3 \cdot 18$ also have this form.
$$5, 5 + 3, 5 + 2 \cdot 3, ..., 5 + 17 \cdot 3$$
also have this form and
$$10, 10 + 3, 10 + 2 \cdot 3, ..., 10 + 16 \cdot 3$$
also have this form.
We are left with 1, 2, 4, 7, 97, 99 and none of them has this form ($19 \cdot 5 < 97$). Hence, in total there are 94 numbers. □

Problem 4.114. *Circles with centers O_1 and O, and respective radii 1 and 5 units, are externally tangent to each other and intersect at point A. The points B and C are points on the circles such that $OB \parallel O_1C$ and the length of the smaller arc AB is equal to the smaller arc AC. What is the angular measure of $\angle AOB$ in degrees?*

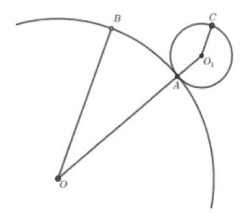

(A) 60 (B) 90 (C) 45 (D) 30 (E) 15

Solution. Answer. (D)
Given that
$$\angle AO_1C = 5 \cdot \angle AOB,$$
and
$$\angle AOB + \angle AO_1C = 180°.$$
Thus, it follows that $\angle AOB = 30°$. □

Problem 4.115. *How many rational numbers $\dfrac{m}{n}$ less than 1 exist, where m and n are positive integers and $\dfrac{m-5}{n-5} = \dfrac{m^2}{n^2}$?*

(A) 0 (B) 1 (C) 2 (D) 3 (E) 4

Solution. Answer. (B)
Given that $m < n$ and
$$(m-5)n^2 = m^2(n-5).$$
Thus, it follows that
$$(mn - 5(m+n))(n-m) = 0.$$
Therefore
$$(m-5)(n-5) = 25.$$
Hence, we obtain that $m - 5 = 1$ and $n - 5 = 25$.
We obtain that $m = 6$, $n = 30$ is the only solution. □

Problem 4.116. *For positive numbers x and y, $y = x^3 - 2x + 2$ and $x = y^3 - 2y + 2$. What is the value of $x + y$?*

(A) 1   (B) 3   (C) 4   (D) 1.5   (E) 2

Solution. Answer. (E)
Given that
$$x + y = x^3 - 2x + 2 + y^3 - 2y + 2.$$
Thus, it follows that
$$x^3 - 3x + y^3 - 3y + 2 = 0.$$
Therefore
$$(x - 1)^2(x + 2) + (y - 1)^2(y + 2) = 0.$$
As
$$x + 2 > 0, y + 2 > 0, (x - 1)^2 \geq 0,$$
and
$$(y - 1)^2 \geq 0,$$
then $x - 1 = 0$ and $y - 1 = 0$.
Hence, we obtain that $x + y = 2$. $\square$

Problem 4.117. *What is the area of the figure consisting of all the points (x, y) in the coordinate plane, where $|x + 2| + |y + 3| \leq 5$ for each such point?*

(A) 25 π (B) 100   (C) 36 π (D) 49 π (E) 50

Solution. Answer. (E)
The figure below is the union of four identical right angle triangles $\varphi_1, \varphi_2, \varphi_3,$ and φ_4.

$$\begin{cases} x + 2 \geq 0, \\ y + 3 \geq 0, \\ x + 2 + y + 3 \leq 5. \end{cases}$$
$(x, y) \in \varphi_1$

$$\begin{cases} x + 2 \geq 0, \\ y + 3 \leq 0, \\ x + 2 - y - 3 \leq 5. \end{cases}$$
$(x, y) \in \varphi_2$

$$\begin{cases} x + 2 \leq 0, \\ y + 3 \geq 0, \\ -x - 2 + y + 3 \leq 5. \end{cases}$$
$(x, y) \in \varphi_3$

$$\begin{cases} x + 2 \leq 0, \\ y + 3 \leq 0, \\ -x - 2 - y - 3 \leq 5. \end{cases}$$
$(x, y) \in \varphi_4$

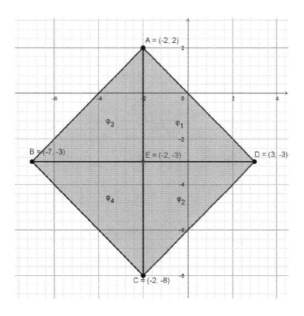

Thus, the figure is a square with diameter 10 and its area is $\frac{10 \cdot 10}{2} = 50$ square units. □

Problem 4.118. *What is the value of the sum of all the two-digit numbers that do not have 1 as a digit?*

(A) 4905 (B) 4000 (C) 4112 (D) 4312 (E) 1260

Solution. Answer. (D)
Note that 9 numbers from those 2-digit numbers start with 2, 3,... and 9. On the other hand, 8 numbers end with 2,3, ..., and 9. Thus, the required sum is

$$9(20 + 30 + 40 + 50 + 60 + 70 + 80 + 90) + 8(2 + 3 + 4 + 5 + 6 + 7 + 8 + 9) = 4312.$$

□

Problem 4.119. *In triangle ABC, $\angle ACB = 120°$ and $AC : CB = 1 : 2$. The points D and E lie on side AB such that $\angle ACD = \angle BCE = 30°$. What is the ratio $DE : AB$?*

(A) 3:10 (B) 3:5 (C) 1:2 (D) 2:3 (E) 2:5

Solution. Answer. (A)
Let $AC = b, CD = x, CE = y$. Thus, it follows that $BC = 2b$.

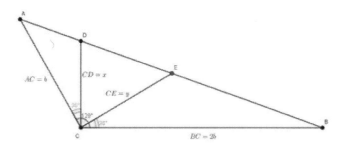

Note that
$$Area(\triangle ADC) + Area(\triangle CDE) = Area(\triangle ACE),$$

127

and
$$Area(\triangle CDE) + Area(\triangle BEC) = Area(\triangle BCD).$$

Hence, we obtain that
$$\frac{1}{2}xb\sin 30° + \frac{1}{2}xy\sin 60° = \frac{1}{2}by,$$

and
$$\frac{1}{2}xy\sin 60° + \frac{1}{2}y \cdot 2b\sin 30° = \frac{1}{2}x \cdot 2b.$$

Therefore
$$\sqrt{3}xy = b(2y - x),$$

and
$$\sqrt{3}xy = b(4x - 2y).$$

We deduce that
$$2y - x = 4x - 2y.$$

Thus, it follows that
$$y = \frac{5}{4}x.$$

Hence
$$x = \frac{2\sqrt{3}}{5}b,$$

and
$$y = \frac{\sqrt{3}}{2}b.$$

We obtain that
$$\frac{DE}{AB} = \frac{Area(\triangle CDE)}{Area(\triangle CAB)} = \frac{xy}{b \cdot 2b} = \frac{\frac{3}{5}b^2}{2b^2} = \frac{3}{10}.$$

□

Problem 4.120. *The lengths of the sides of a rectangular prism are positive integers. The total sum of the numerical values of its volume, total surface area, and the sum of the lengths of all its sides is 2015. What is the volume of the rectangular prism?*

(A) 1000 (B) 1225 (C) 1125 (D) 500 (E) 1200

Solution. Answer. (C)

Let the lengths of the sides of the rectangular prism be a, b, and c, where $a \leq b \leq c$. Then the volume, the total surface area, and the sum of the lengths of all sides are $abc, 2ab + 2ac + 2bc$ and $4a + 4b + 4c$, respectively.

Given that
$$abc + 2ab + 2ac + 2bc + 4a + 4b + 4c = 2015.$$

Thus, it follows that
$$(a+2)(b+2)(c+2) = 2015 + 8 = 2023 = 7 \cdot 17 \cdot 17.$$

Hence, we obtain that
$$a = 5, b = c = 15.$$

Therefore $abc = 1125$.

□

Problem 4.121. Let $ABCD$ be a tetrahedron, such that $AB = 8, AC = 4, AD = 4, BC = \sqrt{34}, BD = 4\sqrt{3}$ and $CD = 5$. What is the volume of the tetrahedron?

(A) $\dfrac{128}{3}$ (B) $\sqrt{39}$ (C) $\dfrac{128}{6}$ (D) $8\sqrt{34}$ (E) $2\sqrt{39}$

Solution. Answer. (E)
Let M be the midpoint of the segment AB.

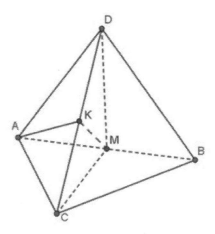

From the formula of the median for the triangles ABC and ABD and medians CM and DM, it follows that
$$CM = \sqrt{\frac{2 \cdot 4^2 + 2 \cdot (\sqrt{34})^2 - 8^2}{4}} = 3,$$
and
$$DM = \sqrt{\frac{2 \cdot 4^2 + 2 \cdot (4\sqrt{3})^2 - 8^2}{4}} = 4.$$

Note that
$$CM^2 + DM^2 = 3^2 + 4^2 = CD^2.$$

Therefore $\angle CMD = 90°$.
Let us calculate the volume of the tetrahedron $AMCD$. As
$$AM = AC = AD,$$

then if K is the midpoint of the line segment CD, we have that $AK \perp CD$. On the other hand, we have that $\triangle AKM$ is congruent to $\triangle AKC$. Hence, we obtain that $AK \perp MK$. Thus, AK is the height of the tetrahedron $AMCD$. Therefore

$$Volume(AMCD) = \frac{1}{3} \cdot \frac{3 \cdot 4}{2} \cdot \sqrt{4^2 - \left(\frac{5}{2}\right)^2} = \sqrt{39}.$$

M is the midpoint of AB and belongs to the plane (MCD), then points A and B are equidistant from that plane. Thus, it follows that
$$Volume(ABCD) = 2\sqrt{39}.$$

□

Problem 4.122. *The first four digits of a twenty-digit number start with 2017, and the other digits are chosen randomly from the digits 0, 1, 2, 3, 4, 5, and 6. What is the probability that the twenty-digit number is divisible by 140?*

(A) $\dfrac{18}{343}$ (B) $\dfrac{15}{343}$ (C) $\dfrac{12}{343}$ (D) $\dfrac{1}{49}$ (E) $\dfrac{4}{343}$

Solution. Answer. (E)
Let us numerate by 1st, 2nd,..., 20th the digits from the leftmost to the rightmost. We have to choose the digits from the 5th to the 20th. In order this number to be divisible by 140, it must be divisible by 5, 4, and 7. It is obvious that the rightmost (20th) digit must be 0 and 19th digit must be one of the numbers from 0, 2, 4, 6. Note that we can choose the digits from the 5th to the 17th arbitrarily, while the 18th digit can be chosen uniquely. This is due to the fact that for any positive number M, numbers

$$100 \cdot 0 + M, 100 \cdot 1 + M, 100 \cdot 2 + M, 100 \cdot 3 + M, 100 \cdot 4 + M, 100 \cdot 5 + M, 100 \cdot 6 + M$$

leave different remainders after division by 7. Therefore, only one of these numbers is divisible by 7. Hence, the required probability is

$$\frac{1}{7} \cdot \frac{4}{7} \cdot \frac{1}{7} = \frac{4}{343}.$$

□

Problem 4.123. *For how many values of a in the interval $(-1, 1)$ does the quadratic $x^2 + ax + 3a + 2$ have at least one integer root?*

(A) 2   (B) 5   (C) 101   (D) 0   (E) 1

Solution. Answer. (A)
Let $a \in (-1, 1)$ and m is an integer root of the quadratic equation

$$x^2 + ax + 3a + 2 = 0.$$

Thus, it follows that
$$a = -\frac{m^2 + 2}{m + 3},$$

and
$$\left|-\frac{m^2 + 2}{m + 3}\right| < 1.$$

Thus, it follows that
$$m^2 + 2 < |m + 3|.$$

Therefore
$$m^2 + 2 < |m| + 3.$$

Hence, we obtain that
$$(2|m| - 1)^2 < 5.$$

We deduce that $m = \{0, 1\}$. Thus, it follows that

$$a \in \{-\frac{3}{4}, -\frac{2}{3}\}.$$

□

Problem 4.124. *Let ABCD be a convex quadrilateral, such that $AB = 1, BC = 4, CD = 8$ and $AD = 7$. What is the greatest possible area of ABCD?*

(A) $2\sqrt{65}$ (B) 17   (C) 18   (D) 19   (E) 18.5

Solution. Answer. (C)
Let us consider point B_1 symmetric to the point B with respect to the perpendicular bisector of the line segment AC.

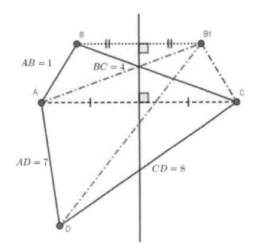

Note that
$$Area(\triangle ABC) = Area(\triangle AB_1C).$$
Thus, it follows that
$$Area(ABCD) = Area(AB_1CD) = Area(\triangle AB_1D) + Area(\triangle B_1CD) \leq$$
$$\leq \frac{1}{2} AB_1 \cdot AD + \frac{1}{2} B_1C \cdot CD = 14 + 4 = 18.$$
Now, let us consider the quadrilateral AB_1CD, where
$$AB_1 = 4, B_1C = 1, CD = 8, AD = 7,$$
and $B_1D = \sqrt{65}$.
In this case, obviously $Area(AB_1CD) = 18$ and $Area(ABCD) = 18$, where B is the symmetric to point B_1 with respect to the perpendicular bisector of the line segment AC.
Therefore, the greatest possible value of the area of $ABCD$ is equal to 18. $\square$

Problem 4.125. *Let ABC be an equilateral triangle with a side length of 2. Let M and N be randomly chosen points on line segments AB and AC, respectively. What is the probability that $MN \leq \sqrt{3}$?*

(A) $\dfrac{\pi}{6}$ (B) $\dfrac{1}{2}$ (C) $\dfrac{1}{3}$ (D) $\dfrac{\pi\sqrt{3}}{6}$ (E) $\dfrac{6-\pi}{\pi}$

Solution. Answer. (D)
Let $MA = u$ and $NA = v$. From the law of cosines, it follows that
$$MN^2 = u^2 - uv + v^2.$$
Therefore
$$u^2 - uv + v^2 \leq 3.$$

131

Hence, we obtain that
$$\left(u - \frac{v}{2}\right)^2 + \left(\frac{v\sqrt{3}}{2}\right)^2 \leq 3.$$
Let us denote
$$u - \frac{v}{2} = x,$$
and
$$\frac{v\sqrt{3}}{2} = y,$$
where $0 \leq u \leq 2$ and $0 \leq v \leq 2$. Therefore
$$0 \leq y \leq \sqrt{3},$$
and
$$0 \leq x + \frac{y}{\sqrt{3}} \leq 2.$$
Hence, we obtain that
$$x^2 + y^2 \leq 3.$$

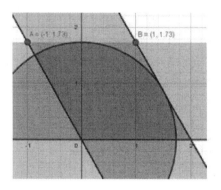

Therefore, the required probability is
$$\frac{\frac{1}{3}\pi(\sqrt{3})^2}{2\sqrt{3}} = \frac{\pi\sqrt{3}}{6}.$$

4.6 Solutions of AMC 10 type practice test 6

Problem 4.126. *What is the value of the following expression*
$$\frac{1}{2} + \frac{1}{2}\cdot\frac{1}{4} + \frac{1}{2}\cdot\frac{3}{4}\cdot\frac{1}{6} + \frac{1}{2}\cdot\frac{3}{4}\cdot\frac{5}{6}\cdot\frac{1}{8} + \frac{1}{2}\cdot\frac{3}{4}\cdot\frac{5}{6}\cdot\frac{7}{8}\cdot\frac{1}{10} - 1?$$

(A) $\dfrac{63}{256}$ (B) $\dfrac{1}{256}$ (C) $-\dfrac{63}{256}$ (D) $-\dfrac{1}{256}$ (E) $\dfrac{65}{256}$

Solution. Answer. (C)
We have that
$$\frac{1}{2} + \frac{1}{2}\cdot\frac{1}{4} + \frac{1}{2}\cdot\frac{3}{4}\cdot\frac{1}{6} + \frac{1}{2}\cdot\frac{3}{4}\cdot\frac{5}{6}\cdot\frac{1}{8} + \frac{1}{2}\cdot\frac{3}{4}\cdot\frac{5}{6}\cdot\frac{7}{8}\cdot\frac{1}{10} - 1 =$$
$$\frac{1}{2}\cdot\frac{1}{4} - \frac{1}{2} + \frac{1}{2}\cdot\frac{3}{4}\cdot\frac{1}{6} + \frac{1}{2}\cdot\frac{3}{4}\cdot\frac{5}{6}\cdot\frac{1}{8} + \frac{1}{2}\cdot\frac{3}{4}\cdot\frac{5}{6}\cdot\frac{7}{8}\cdot\frac{1}{10} =$$
$$\frac{1}{2}\cdot\frac{3}{4}\cdot\frac{1}{6} - \frac{1}{2}\cdot\frac{3}{4} + \frac{1}{2}\cdot\frac{3}{4}\cdot\frac{5}{6}\cdot\frac{1}{8} + \frac{1}{2}\cdot\frac{3}{4}\cdot\frac{5}{6}\cdot\frac{7}{8}\cdot\frac{1}{10} =$$
$$\frac{1}{2}\cdot\frac{3}{4}\cdot\frac{5}{6}\cdot\frac{1}{8} - \frac{1}{2}\cdot\frac{3}{4}\cdot\frac{5}{6} + \frac{1}{2}\cdot\frac{3}{4}\cdot\frac{5}{6}\cdot\frac{7}{8}\cdot\frac{1}{10} =$$
$$\frac{1}{2}\cdot\frac{3}{4}\cdot\frac{5}{6}\cdot\frac{7}{8}\cdot\frac{1}{10} - \frac{1}{2}\cdot\frac{3}{4}\cdot\frac{5}{6}\cdot\frac{7}{8} = -\frac{1}{2}\cdot\frac{3}{4}\cdot\frac{5}{6}\cdot\frac{7}{8}\cdot\frac{9}{10} = -\frac{63}{256}.$$

Alternative solution. We have that
$$\frac{128 + 32 + 16 + 10 + 7 - 256}{256} = -\frac{63}{256}.$$

Problem 4.127. *Jerry solved 100 problems in n days. Each day, he solved either 5, 6 or 7 problems. What is the smallest possible value of n?*

(A) 14 (B) 15 (C) 20 (D) 16 (E) 17

Solution. Answer. (B)
Note that
$$\frac{100}{7} > 14.$$
Thus, it follows that $n \geq 15$. We have that
$$12\cdot 7 + 1\cdot 6 + 2\cdot 5 = 100.$$
Therefore, the possible minimum value of n is 15.

Problem 4.128. *In a rectangular room with sides 5m and 6m, there are: a bookshelf, a table, a sofa, and an armchair. They respectively occupy rectangular spaces of size $0.75 \times 2.0m$, $0.8 \times 4.0m$, $0.9 \times 2.0m$, and $0.9 \times 1.0m$. What part of the room is empty?*

(A) $\dfrac{37}{150}$ (B) $\dfrac{1}{2}$ (C) $\dfrac{1}{3}$ (D) $\dfrac{113}{150}$ (E) $\dfrac{2}{3}$

Solution. Answer. (D)
The floor area of the room is 30, and the occupied area is
$$1.5 + 3.2 + 1.8 + 0.9 = 7.4.$$
Therefore, the area of the floor that is still not occupied is $30 - 7.4 = 22.6$. We obtain that
$$\frac{22.6}{30} = \frac{113}{150}$$
is the part of the base that is not occupied.

Problem 4.129. *At 9 : 00 AM, two snails start moving simultaneously from the same point in different directions on a circular track with a circumference of 21 meters. Every hour, they change the direction in which they move. They cover a distance of $2 - (-1)^n$ m and $4 - 2(-1)^n$ m after the n th hour. At what time will the snails meet each other for the first time?*

(A) 16:00   (B) 15:20   (C) 15:30   (D) 15:00   (E) 14:00

Solution. Answer. (E)
The first one covers 2m distance in 2 hours, 4m in 4 hours and for the 1st time covers 7m in 5hours. On the other hand, the second one covers 14m in 5 hours. Therefore, the answer is 14 : 00. □

Problem 4.130. *The sum of the squares of two positive numbers is three times the product of the numbers. How many times larger is the square of the sum of these numbers than the square of their difference?*

(A) 5   (B) 3   (C) 9   (D) 2   (E) 4

Solution. Answer. (A)
Let us assume that these positive numbers are a and b. Given that
$$a^2 + b^2 = 3ab.$$
Thus, it follows that
$$\frac{(a+b)^2}{(a-b)^2} = \frac{a^2 + b^2 + 2ab}{a^2 + b^2 - 2ab} = \frac{5ab}{ab} = 5.$$
□

Problem 4.131. *A student was assigned a test consisting of 10 problems. Each correct answer received 2 points, an incomplete answer received 1 point, and no points were given for a wrong answer. The student's total score after the test was 19. Which of the following statements is true:*

(A) The student correctly answered all 10 problems.
(B) At least one of the answers was incorrect.
(C) All the answers for all 10 problems were incomplete.
(D) Only one answer was correct.
(E) Only one answer was incomplete.

Solution. Answer. (E)
(A) is not true as in this case the total score is $10 \cdot 2 = 20$ points. (B) is not true as student's total score couldn't be more than $9 \cdot 2 = 18$ points. (C) is not true as in this case student's total score is $10 \cdot 1 = 10$ points. (D) is not true as student's total score couldn't be more than $9 \cdot 1 + 2 = 11 points$. (E) is true as $9 \cdot 2 + 1 = 19$. □

Problem 4.132. *By what percent is the circumference of a circle inscribed in a square less than the perimeter of the square?*

(A) 50%   (B) 22%   (C) 21%   (D) $\frac{100(4-\pi)}{\pi}$%   (E) $25(4-\pi)$%

Solution. Answer. (E)
Let the side of the square be a. Thus, it follows that the perimeter of a square is $4a$. The radius of the circle inscribed in that square is $\frac{a}{2}$ and the circumference is πa. Therefore, the required percentage is
$$\frac{4a - \pi a}{4a} \cdot 100\% = 25(4-\pi)\%.$$
□

Problem 4.133. *The participants of a mathematics conference stay in two hotels. Participants staying in the same hotel shook hands with each other, while participants staying in different hotels did not. The total number of handshakes is equal to the product of the number of participants in each hotel. What is the number of conference participants, if it is larger than 17 and less than 34?*

(A) 20   (B) 25   (C) 18   (D) 30   (E) 33

Solution. Answer. (B)
Let m and n be the number of participants staying in those hotels. The total number of handshakes is $\binom{m}{2} + \binom{n}{2} = \frac{m(m-1)}{2} + \frac{n(n-1)}{2}$. Given that $\frac{m(m-1)}{2} + \frac{n(n-1)}{2} = mn$. Thus, it follows that $m + n = (m-n)^2$ and $17 < m + n < 34$. Hence, we obtain that $m + n = 25$. □

Problem 4.134. *A boat starts moving downstream, and covers a distance equal to the distance that would have taken three hours if it were moving upstream. Then it moves upstream. It travels a distance moving upstream equal to the distance that would have taken two hours if it were moving downstream. What is the ratio of the boat's rate to the rate of the current, if the boat's whole trip took 5 hours?*

(A) 5   (B) 4   (C) 10   (D) 2   (E) 3

Solution. Answer. (A)
Let us denote the rate of the boat by x km/h and the current rate by y km/h. The boat moved downstream for $\frac{3(x-y)}{x+y}$ hours, and upstream for $\frac{2(x+y)}{x-y}$ hours. Given that

$$\frac{3(x-y)}{x+y} + \frac{2(x+y)}{x-y} = 5.$$

Thus, it follows that

$$2\left(\frac{x+y}{x-y} - 1\right) = 3\left(1 - \frac{x-y}{x+y}\right).$$

Therefore
$$2(x+y) = 3(x-y).$$

Hence, we obtain that $\frac{x}{y} = 5$. □

Problem 4.135. *Square ABCD has a side length of 12. M is an interior point of ABCD, and the distance from M to sides AB, AD, and CD is a, b, and c, respectively. How many possible points M are there, such that a, b, and c are integers and there exists a triangle with sides a, b, and c?*

(A) 72   (B) 60   (C) 66   (D) 59   (E) 61

Solution. Answer. (E)
All such points are drawn in the figure below and their total number is

$$2(1 + 3 + 5 + 7 + 9 + 11) - 11 = 61.$$

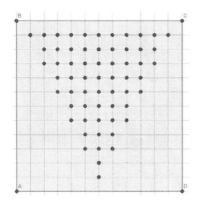

Problem 4.136. *Given a cube with a side length of 1 unit. Let Φ be the solid formed by all points that are located within a distance of 1 unit from any point on the surface of the cube. What is the volume of Φ?*

(A) 27　　(B) $7 + 3\pi$　　(C) $7 + \dfrac{4\pi}{3}$　　(D) $7 + \dfrac{13\pi}{3}$　　(E) $6 + 4\pi$

Solution. Answer. (D)
The solid Φ consists of 7 unit cubes, 12 solids each 4 of which create a cylinder of radius 1 and height 1 and also 8 solids which create a sphere of radius 1. Therefore, the volume of solid Φ is
$$7 + 3\pi + \frac{4\pi}{3} = 7 + \frac{13\pi}{3}.$$

□

Problem 4.137. *What is the loci of all points (x, y) for which the inequality $|x + y - 1| + |x - y + 1| + |x + y + 1| + |x - y - 1| \leq 4$ holds true?*

(A) four vertices of a square　　(B) a square and its inner region　　(C) a triangle and its inner region
(D) three vertices of a triangle　　(E) eight points

Solution. Answer. (B)
Let that image be φ. It is known that
$$|a| + |b| \geq |a + b|.$$
Morever
$$|a| + |b| = |a + b|,$$
when $ab \geq 0$. According to this inequality, we have that
$$4 \geq |x + y - 1| + |x - y - 1| + |x + y + 1| + |x - y + 1| \geq |2x - 2| + |2x + 2| = |2 - 2x| + |2x + 2| \geq 4.$$
Thus, it follows that
$$|2 - 2x| + |2 + 2x| = 4, |1 - x - y| + |1 - x + y| = |2 - 2x|, |1 + x + y| + |1 + x - y| = |2 + 2x|.$$
Therefore φ is the loci of all (x, y), such that
$$\begin{cases} 2 - 2x \geq 0, \\ 2 + 2x \geq 0, \\ 1 - x - y \geq 0, \\ 1 - x + y \geq 0, \\ 1 + x + y \geq 0, \\ 1 + x - y \geq 0. \end{cases}$$

Hence, the answer is (B).

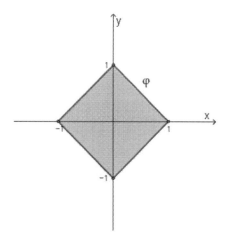

Problem 4.138. *Let the number sequence x_n be the remainder when $x_{n-1}^2 + x_{n-2}^3$ is divided by 5, for n = 3, 4, ... and $x_1 = 1, x_2 = 2$. What is x_{2017}?*

(A) 0   (B) 1   (C) 2   (D) 3   (E) 4

Solution. Answer. (B)
Given that
$$x_1 = 1, x_2 = 2, x_3 = 0, x_4 = 3, x_5 = 4, x_6 = 3, x_7 = 3, x_8 = 1, x_9 = 3,$$
$$x_{10} = 0, x_{11} = 2, x_{12} = 4, x_{13} = 4, x_{14} = 0, x_{15} = 4, x_{16} = 1,$$
$$x_{17} = 0, x_{18} = 1, x_{19} = 1, x_{20} = 2, ...$$
Therefore, the sequence x_n is periodic and the period is 18. Thus, it follows that
$$x_{2017} = x_{1+2016} = x_{1+18 \cdot 112} = x_1 = 1.$$

Problem 4.139. *If Pablo gives 20% of his money to Mary, then Mary would have 25% more money than Pablo. If Mary gives 20% of her money to Pablo, he would have more money than Mary by what percent?*

(A) 81.25%   (B) 52%   (C) $44\frac{24}{29}\%$ (D) $57\frac{11}{17}\%$ (E) 25%

Solution. Answer. (A)
Let Pablo has 100 units of money, and Mary has x units of money. Given that
$$x + 20 = 80 \cdot \frac{125}{100}.$$
Thus, it follows that $x = 80$. We need to find how many percent more than 64 is
$$100 + \frac{80 \cdot 20}{100}.$$
The required percent is
$$\frac{116 - 64}{64} \cdot 100\% = 81,25\%.$$

137

Problem 4.140. *Circles with radii 8 and 2 are positioned such that the distance between their centers is 3. A circle with radius 1 is randomly placed within the circle with radius 8. What is the probability that the circles with radii 1 and 2 intersect?*

(A) $\dfrac{1}{16}$ (B) $\dfrac{9}{64}$ (C) $\dfrac{9}{49}$ (D) $\dfrac{1}{5}$ (E) $\dfrac{1}{2}$

Solution. Answer. (C)

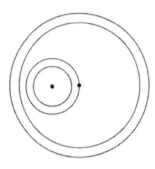

Note that in order the circle of radius 1 to be in the circle of radius 8 its center must lie in the circle of radius 7 (see the figure). On the other hand, in order the circles with radii 1 and 2 to have a common point the center of the circle of radius 1 must lie in the circle of radius 3. Therefore, the required probability is $\dfrac{9}{49}$. □

Problem 4.141. *A natural number is considered "nice", if at least six of its divisors are from the set of $\{1, 2, 3, 4, 5, 6, 7, 8, 9, 10\}$. What is the smallest possible value of the positive difference of two "nice" numbers?*

(A) 1 (B) 3 (C) 6 (D) 4 (E) 2

Solution. Answer. (E)
Let a and b be "nice" numbers, such that $a > b$. Note that both a and b have at least two common divisors. Therefore, there exists a positive integer c greater than 1, such that $c \mid a$ and $c \mid b$. Thus, it follows that $c \mid (a - b)$. Therefore $a - b \geq c \geq 2$. We obtain that $a - b \geq 2$. Note that 882 and 880 are "nice" numbers and their positive difference is 2. □

Problem 4.142. *All the points with integer coordinates (x, y) on the circle given by $x^2 + y^2 = 65$ form a polygon. What is the area of the polygon?*

(A) 128 (B) 198 (C) 256 (D) 200 (E) 70

Solution. Answer. (B)
The points with integer coordinates on the circle are (1,8), (4,7), (7,4), (8,1), (-1,8), (-4,7), (-7,4), (-8,1), (-1, -8), (-4, -7), (-7, -4), (-8, -1), (1, -8), (4, -7), (7, -4), (8, -1).

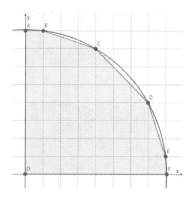

Note that the the required area is four times the area of polygon $OABCDEF$, that is

$$4\left(8\cdot 8 - 2\frac{1\cdot 3}{2} - \frac{3\cdot 3}{2} - 2\cdot 1\cdot 3 - 1\cdot 1\right) = 198.$$

□

Problem 4.143. *Mary is randomly choosing three numbers from the set of 1, 2, ..., 10. What is the probability of the event that the chosen numbers form either an arithmetic or a geometric progression?*

(A) $\dfrac{23}{120}$ (B) $\dfrac{1}{6}$ (C) $\dfrac{5}{24}$ (D) $\dfrac{1}{5}$ (E) $\dfrac{1}{4}$

Solution. Answer. (D)
The number of chosen triples is

$$\binom{10}{3} = 120.$$

Let us place the chosen numbers in increasing order. The numbers of arithmetic sequences with differences 1, 2, 3, 4 are 8, 6, 4, 2, respectively. On the other hand, the geometric sequences are four: either 1, 2, 4 or 1, 3, 9 or 2, 4, 8 or 4, 6, 9. The required probability is

$$\frac{24}{120} = \frac{1}{5}.$$

□

Problem 4.144. *Five chairs are placed around a circular table. In how many different ways can two girls and three boys be seated, such that two girls do not sit next to each other?*

(A) 120 (B) 60 (C) 30 (D) 20 (E) 24

Solution. Answer. (B)
First girl can choose a seat in 5 ways, the second girl has two choices. Note that once girls are done with the choice of their seats then the boys can choose their seats in 3! ways. Therefore, the answer is $5\cdot 2\cdot 3! = 60$.

□

Problem 4.145. Let $n = \overline{a_1 a_2 \ldots a_k}$ and $T(n) = |a_1 - a_2 + \ldots + (-1)^{k-1} a_k|$. For example $T(1237) = |1-2+3-7| = 5$. For some natural number n, $T(n) = 4$. Which of the following values can be $T(n-1)$?

(A) 2   (B) 9   (C) 6   (D) 1   (E) 7

Solution. Answer. (C)
Note that
$$T(n) = |a_1 - a_2 + \ldots + (-1)^{k-1} a_k| = |(-1)^{k-1} a_1 - (-1)^{k-2} a_2 + \ldots + a_k|,$$
and
$$\overline{a_1 a_2 \ldots a_k} - ((-1)^{k-1} a_1 - (-1)^{k-2} a_2 + \ldots + a_k) = a_1(10^{k-1} - (-1)^{k-1}) + a_2(10^{k-2} - (-1)^{k-2}) + \ldots + a_{k-1}(10-(-1)).$$
Obviously
$$11 \mid 10^m - (-1)^m,$$
for any positive integer m. Thus, it follows that
$$11 \mid n - T(n),$$
or
$$11 \mid n + T(n).$$
If $T(n)$ leaves a remainder of 4 after division by 11, then $n = 11k + 4$ or $n = 11k + 7$, for some positive integer k. Therefore
$$n - 1 = 11k + 3,$$
or
$$n - 1 = 11k + 6.$$
Note that $n-1$ can leave remainders 3, 8, 6 or 5 after division by 11. Hence, we obtain that $T(n-1)$ can leave only a remainder of 6 after division by 11.
Example. If $n = 3010$, then $T(n-1) = 6$. □

Problem 4.146. *A square with a side length of x is inscribed into a triangle with side lengths of 13, 14, and 15, such that two of its vertices lie on the smallest side of the triangle. Another square, with a side length of y, is inscribed into a second congruent triangle, such that two of its vertices lie on the biggest side of the triangle. What is $\dfrac{15}{x} - \dfrac{13}{y}$?*

(A) $\dfrac{56}{195}$ (B) 1   (C) $\dfrac{1685}{1703}$ (D) $\dfrac{2}{3}$ (E) $\dfrac{3}{2}$

Solution. Answer. (A)

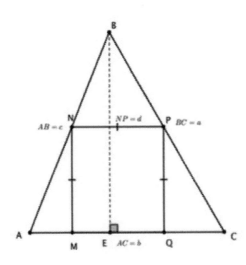

Let $MNPQ$ be the square inscribed in ABC. Given that
$$AB = c, BC = a, AC = b, MN = d.$$
Let $BE \perp AC, BE = h_b$. We have that $NP \parallel AC$. Thus, it follows that
$$\angle BNP = \angle BAC.$$
Therefore
$$\triangle BNP \sim \triangle BAC.$$
Hence, we obtain that
$$\frac{h_b - d}{h_b} = \frac{d}{b}.$$
We deduce that
$$d = \frac{2S}{b + h_b},$$
where S is the area of triangle ABC. Thus, it follows that
$$d = \frac{2Sb}{b^2 + 2S}.$$
Taking this into consideration, we obtain that
$$\frac{15}{x} - \frac{13}{y} = 15\left(\frac{13}{2S} + \frac{1}{13}\right) - 13\left(\frac{15}{2S} + \frac{1}{15}\right) = \frac{15}{13} - \frac{13}{15} = \frac{56}{195}.$$
$\square$

Problem 4.147. *Equilateral triangle ABC is inscribed in a circle with center O and radius R. A circle σ, with center O and radius $\frac{R\sqrt{3}}{3}$, is constructed. What is the area of the portion of triangle ABC that lies outside circle σ?*

(A) $\dfrac{R^2\sqrt{3}}{3}$ (B) $\dfrac{3\sqrt{3} - \pi}{6}R^2$ (C) $\dfrac{3\sqrt{3} + 2\pi}{6}R^2$ (D) $\dfrac{4 - \pi}{6}R^2$ (E) $\dfrac{2\sqrt{3} - \pi}{3}R^2$

Solution. Answer. (B)

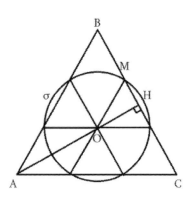

Given that
$$OM = \frac{\sqrt{3}R}{3},$$
and $OH \perp BC$. We have that
$$OH = \frac{R}{2}.$$

Thus, it follows that
$$MH = \sqrt{\frac{R^2}{3} - \frac{R^2}{4}} = \frac{\sqrt{3}R}{6} = \frac{OM}{2}$$
and $\angle MOH = 30°$. Therefore
$$\angle OMH = 60° = \angle ABC.$$

Hence, the required area is
$$\frac{\sqrt{3}}{4} \cdot (\sqrt{3}R)^2 - 3 \cdot \frac{1}{9} \cdot \frac{\sqrt{3}}{4} \cdot (\sqrt{3}R)^2 - \frac{1}{2} \cdot \pi \left(\frac{\sqrt{3}R}{3}\right)^2 = \frac{3\sqrt{3} - \pi}{6} R^2.$$
$\square$

Problem 4.148. *Points with integer coordinates (i, j), where i and j are between 1 and 4 inclusive, are placed on the coordinate plane. Out of these 16 points, how many points, at most, can belong to a circle?*

(A) 3   (B) 4   (C) 5   (D) 8   (E) 10

Solution. Answer. (D)
Note that these 16 points lie on four lines and they might have at most two intersections with the circle. Thus, from given 16 points not more than 8 are on the same circle. Note that points (1,2), (1,3), (2,4), (3,4), (4,3), (4,2), (2,1), (3,1) are on the same circle. Therefore, the answer is 8. $\square$

Problem 4.149. *There exists a rational number k, such that each of the polynomials $x^3 + x^2 + kx + 2$ and $x^4 - 4x^3 + 4x^2 + (k+5)x - 3$ has an integer root larger than 1. What is k?*

(A) $\frac{5}{24}$ (B) $-\frac{3}{25}$ (C) -7   (D) 70   (E) 3

Solution. Answer. (C)
Let m and n be the roots of those polynomials greater than 1, in that case
$$k = -m^2 - m - \frac{2}{m},$$
and
$$k + 5 = -n^3 + 4n^2 - 4n + 3n.$$
Thus, it follows that
$$\frac{3}{n} + \frac{2}{m} = n^3 - 4n^2 + 4n - m^2 - m + 5,$$
is an integer, except the case when
$$\frac{3}{n} + \frac{2}{m} \leq \frac{3}{2} + 1 = 2.5.$$
Hence, we obtain that
$$\frac{3}{n} + \frac{2}{m} = 1,$$
or
$$\frac{3}{n} + \frac{2}{m} = 2.$$
We deduce that
$$(n-3)(m-2) = 6,$$
or
$$(2n-3)(m-1) = 3.$$

Therefore, the possible values are $m = 3, n = 9$ or $m = 4, n = 6$ or $m = 5, n = 5$ or $m = 8, n = 4$ or $m = 2, n = 3$ or $m = 4, n = 2$.

Only $m = 2, n = 3$ satisfies the following equation
$$\frac{3}{n} + \frac{2}{m} = n^3 - 4n^2 + 4n - m^2 - m + 5.$$

Thus, it follows that $k = -7$. $\square$

Problem 4.150. *A three-digit number is considered "fancy" if none of its digits exceed 8, and if increasing each digit by 1 results in a new three-digit number that is divisible by 11. How many possible three-digit "fancy" numbers are there?*

(A) 81 (B) 55 (C) 60 (D) 54 (E) 56

Solution. Answer. (E)

Let $\overline{abc}$ be a "fancy" number. Given that
$$11 \mid \overline{(a+1)(b+1)(c+1)} = 99(a+1) + 11(b+1) + a - b + c + 1,$$

or $11 \mid a + c - b + 1$. Thus, it follows that $a + c - b = -1$ or $a + c - b = 10$. Hence, we obtain that $a + c = b - 1$ or $a + c = b + 10$, where $a = \overline{1,8}$ and $b, c = \overline{0,8}$.

The total number of "fancy" numbers satisfying $a + c = b - 1$ is $1 + 2 + ... + 7 = 28$.
The total number of "fancy" numbers satisfying $a + c = b + 10$ is again $1 + 2 + ... + 7 = 28$.
Therefore, the total number of "fancy" numbers is 56. $\square$

4.7 Solutions of AMC 10 type practice test 7

Problem 4.151. *What is the value of $2^{3^{10}} - 8^{3^9}$?*

(A) 2   (B) 0   (C) 2020   (D) 1   (E) 1024

Solution. Answer. (B)
We have that
$$2^{3^{10}} = (2^3)^{3^9} = 8^{3^9}.$$
Thus, it follows that
$$2^{3^{10}} - 8^{3^9} = 0.$$

$\square$

Problem 4.152. *How many 0 digits are there at the end of $(18! - 15!)$?*

(A) 2   (B) 3   (C) 4   (D) 5   (E) 1

Solution. Answer. (C)
We have that
$$18! - 15! = 15! \cdot (16 \cdot 17 \cdot 18 - 1) = 15! \cdot 5 \cdot 979 = 5^4 \cdot 2^4 \cdot n,$$
where n is not divisible by 5. Thus, this expression ends in four 0's. $\square$

Problem 4.153. *Anna and Bonita were born on the same date in different years. What is Anna's age, if two years ago she was twice as old as Bonita, and three years ago she was three times elder than Bonita.*

(A) 3   (B) 4   (C) 5   (D) 6   (E) 7

Solution. Answer. (D)
Let us denote the ages of Anna and Bonita by a and b, respectively. Given that
$$a - 2 = 2(b - 2),$$
and
$$a - 3 = 3(b - 3).$$
Thus, it follows that $a = 2b - 2$ and $2b - 2 - 3 = 3(b - 3)$. Hence, we obtain that $b = 4$ and $a = 6$. Therefore $a = 6$. $\square$

Problem 4.154. *The unit squares of a 5×12 grid are colored like a chessboard in alternating black and white colors. The diagonal of the rectangle is drawn. The intersections of the diagonal and the black squares are line segments. What is the total length, in units, of these segments?*

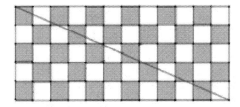

(A) 6.5   (B) 8   (C) 9   (D) 9.5   (E) 10

Solution. Answer. (A)

The common part between the diagonal of the rectangle and any square is either nonexistent or a line segment. In the latter case, let us name the segment with color associated with the color of the square (that is, white segment or black segment). Note that, the symmetric segment of any black segment on the diagonal with respect to the midpoint of the diagonal is a white segment and vice versa. The diagonal has a length 13 and half of it are black segments, so their total length is 6.5. □

Problem 4.155. *Given two congruent circles on the plane with non-coincident centers. How many of the transformations below map one circle onto another?*
- *parallel translation.*
- *point symmetry.*
- *line symmetry.*
- *rotation by 30° angle.*

(A) 4 (B) 3 (C) 2 (D) 1 (E) 0

Solution. Answer. (A)
All 4 possible cases are in drawn below:

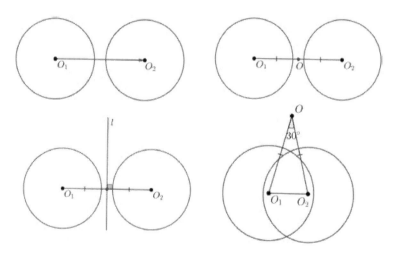

□

Problem 4.156. *What is the greatest integer that cannot be written as a sum of two composite numbers?*

(A) 101 (B) 11 (C) 111 (D) 2019 (E) 42

Solution. Answer. (B)
If n is an even number and $n \geq 8$. Thus, it follows $n = 4 + (n-4)$, where 4 and $n-4$ are composite numbers.
If n is an odd number and $n \geq 13$, we have that $n = 9 + (n-9)$, where 9 and $n-9$ are composite numbers.
Note that 11 cannot be written as a sum of two composite numbers and it is the largest with this property. □

Problem 4.157. *How many odd positive divisors does 30^{10} have?*

(A) 121   (B) 1331   (C) 665   (D) 666   (E) 667

Solution. Answer. (A)
We have that
$$30^{10} = 2^{10} \cdot 3^{10} \cdot 5^{10}.$$
Thus, it follows that any odd divisor of this number is of the form $3^a \cdot 5^b$, where $a, b \in \{0, 1, 2, ..., 10\}$. Therefore, there are $11 \cdot 11 = 121$ odd divisors. $\square$

Problem 4.158. *How many of given quadrilaterals have an interior point that is equidistant from each line containing each side of this quadrilateral?*
- *Square.*
- *Rhombus.*
- *Parallelogram whose adjacent sides are not congruent.*
- *Isosceles trapezoid whose legs are congruent to the midsegment.*
- *Trapezoid for which the sum of the lengths of the legs is more than twice the midsegment.*

(A) 1   (B) 4   (C) 2   (D) 5   (E) 3

Solution. Answer. (E)
Given condition is equivalent to that the circle can be inscribed in the quadrilateral. Note that from the listed geometric figures it is possible to inscribe a circle to a square, rhombus, and isosceles trapezoid, whose legs are congruent to the midsegment. Thus, it follows that the answer is 3. $\square$

Problem 4.159. *A geometric sequence consists of five terms. The arithmetic mean of the first four terms of this sequence is 10. The arithmetic mean of the last four terms is 30. What is the fifth term of the sequence?*

(A) 40   (B) 52   (C) 64   (D) 72   (E) 81

Solution. Answer. (E)
Let the terms of the geometric sequence be b_1, b_2, b_3, b_4 and b_5. Given that
$$\frac{b_1 + b_2 + b_3 + b_4}{4} = 10,$$
and
$$\frac{b_2 + b_3 + b_4 + b_5}{4} = 30.$$
We have that
$$b_2 + b_3 + b_4 + b_4 = (b_1 + b_2 + b_3 + b_4) \cdot r,$$
where r is the common ratio of the geometric sequence. Therefore, $10 \cdot r = 30$ or $r = 3$. Hence, we obtain that
$$b_1 + 3b_1 + 9b_1 + 27b_1 = 40.$$
Therefore $b_1 = 1$. Thus, it follows that
$$b_5 = b_1 \cdot r^4 = 81.$$

$\square$

Problem 4.160. *A shop sells two-colored balls. Ten of the balls in stock are red and blue, 7 are blue and yellow, and 9 are red and yellow. Suppose n balls are chosen at random. What is the least possible value of n such that you can be certain at least 12 balls will have the same color on them?*

(A) 14   (B) 16   (C) 17   (D) 15   (E) 18

Solution. Answer. (C)

Let a, b and c be the quantity of balls taken from 10, 7 and 9 ball sets, respectively. Note that, when

$$a = 5, b = 6, c = 5,$$

then the condition of the problem is not satisfied. Therefore $n > 5 + 6 + 5 = 16$. When 17 balls are chosen, we have that

$$34 = (a + b) + (b + c) + (c + a),$$

so at least one of the summands $a+b, b+c$ or $a+c$ is not less than 12. Thus, the condition of the problem is satisfied. □

Problem 4.161. *Let ABC be a triangle, such that $AC = 10, BC = 17$ and $AB = 21$. Let M and N be points on side AB, such that $CM^2 = AM \cdot BM$ and $CN^2 = AN \cdot BN$ (see the figure). What is the value of the area of triangle CMN?*

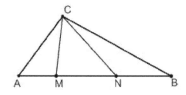

(A) 34   (B) 42   (C) 30   (D) 36   (E) $10\sqrt{3}$

Solution. Answer. (A)

Let us choose the standard rectangular coordinate system as shown in the figure.

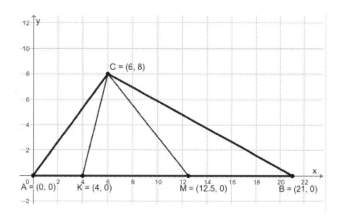

Let $C(x_0, y_0)$. Given that

$$x_0^2 + y_0^2 = 100,$$

and

$$(x_0 - 21)^2 + y_0^2 = 289.$$

Thus, it follows that

$$42x_0 - 441 = 100 - 289,$$

or $x_0 = 6$ and $y_0 = 8$.
Therefore $K(x, 0)$ lies on segment AB, such that
$$CK^2 = AK \cdot BK.$$
Hence
$$(x-6)^2 + 8^2 = x(21-x).$$
We obtain that $x = 4$ or $x = 12.5$. Thus, it follows that $MN = 12.5 - 4 = 8.5$ and
$$Area(CMN) = \frac{8 \cdot 8.5}{2} = 34.$$

Alternative solution. This problem can also be solved using Stewart's theorem. An interested reader can do it independently. □

Problem 4.162. *How many positive integers N not exceeding 2020 have the following property: the sum of the first N positive integers divides the sum of its squares?*

(A) 404 (B) 505 (C) 674 (D) 1010 (E) 1011

Solution. Answer. (C)
Let us use the following two formulas.
$$\begin{cases} 1 + 2 + ... + n = \frac{n(n+1)}{2}, \\ 1^2 + 2^2 + ... + n^2 = \frac{n(n+1)(2n+1)}{6}. \end{cases}$$
Taking this into consideration and from the condition of the problem, we obtain that
$$\frac{n(n+1)}{2} \mid \frac{n(n+1)(2n+1)}{6}.$$
Hence, we deduce that $\frac{2n+1}{3}$ is an integer. Therefore, n leaves a remainder 1 after division by 3. Note that such numbers not exceeding 2020 are 1, 4, ..., 2020 and their total number is
$$\frac{2020-1}{3} + 1 = 674.$$

Alternative solution. We have that
$$(k+1)^3 - k^3 = 3k^2 + 3k + 1.$$
Letting $k = 1, 2, ..., n$ and summing up the equations, we obtain that
$$(n+1)^3 - 1 = 3(1^2 + 2^2 + ... + n^2) + 3(1 + 2 + ... + n) + n.$$
Thus, it follows that
$$\frac{n(n+1)(n+2)}{3} = 1^2 + 2^2 + ... + n^2 + 1 + 2 + ... + n.$$
Therefore, the condition of the problem is satisfied if and only if
$$1 + 2 + ... + n = \frac{n(n+1)}{2} \mid \frac{n(n+1)(n+2)}{3},$$
or equivalently $3 \mid (n+2)$.
Thus, it follows that n leaves a remainder 1 after division by 3. Note that such numbers not exceeding 2020 are 1, 4, ..., 2020 and their total number is
$$\frac{2020-1}{3} + 1 = 674.$$
□

Problem 4.163. *Let points M, N lie on sides BC, CD of parallelogram $ABCD$, respectively. Given that $\angle BAD = 85°, \angle MAN = 20°$ and $\angle MNA = 10°$. Let ME be the angle bisector in $\triangle AMN$. What is the angle measure of $\angle BEC$?*

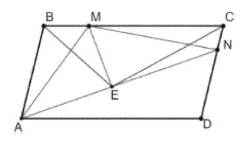

(A) 100   (B) 120   (C) 90   (D) 150   (E) 160

Solution. Answer. (D)

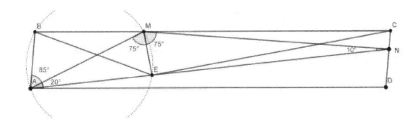

We have that
$$\angle AME = \frac{180° - 20° - 10°}{2} = 75°,$$
while from $\triangle AME$, we get
$$\angle AEM = 180° - 75° - 20° = 85°.$$
We have also that
$$\angle ABC = 180° - 85° = 95°.$$
Therefore
$$\angle ABM + \angle AEM = 180°.$$
Thus, points A, B, M and E are concyclic. Therefore
$$\angle EBM = \angle EAM = 20°.$$
In a similar way, we get that points M, C, N and E are concyclic. Hence
$$\angle ECM = 10°.$$
Thus, it follows that
$$\angle BEC = 180° - 20° - 10° = 150°.$$

□

Problem 4.164. *Consider four congruent circles on a plane, such that they are pairwise non-concentric and they divide the plane in n parts. How many values of n are possible?*

(A) 10   (B) 9   (C) 8   (D) 7   (E) 6

Solution. Answer. (A)

Note that m congruent circles divide a plane into at least $m + 1$ parts as each new circle increases the part by at least 1. On the other hand, these m circles divide the plane into at most $m^2 - m + 2$ parts at m^{th} circle can increase the number of parts by not more than $2m - 2$.

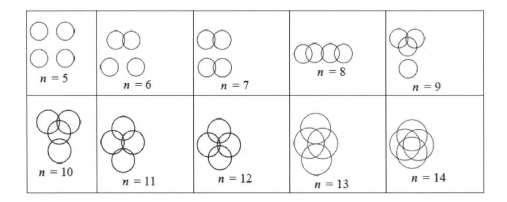

Problem 4.165. *What is the least possible value of the expression $(x+4)(x+5) + \dfrac{10^6}{x(x+9)}$, where x is a positive real number?*

(A) 2017   (B) 2018   (C) 2019   (D) 2020   (E) 2021

Solution. Answer. (D)
Let us do the following substitution
$$x(x+9) = y.$$
Thus, it follows that
$$(x+4)(x+5) + \frac{10^6}{x(x+9)} = y + \frac{10^6}{y} + 20 = \left(\sqrt{y} - \frac{10^3}{\sqrt{y}}\right)^2 + 2020 = (\sqrt{x(x+9)} - \frac{10^3}{\sqrt{x(x+9)}})^2 + 2020.$$

Therefore, from this equation we obtain that the least possible value of the expression $(x+4)(x+5) + \dfrac{10^6}{x(x+9)}$ is 2020. Note that this value is reachable when
$$x(x+9) = 10^3,$$
or
$$x = \frac{-9 + \sqrt{4081}}{2}.$$

Problem 4.166. *Let x and y be randomly chosen numbers from $[0, 1]$. What is the probability that the following inequality holds true?*
$$|x - 0.5| + |y - 1.5| \leq 1.$$

(A) $\dfrac{1}{4}$ (B) $\dfrac{3}{4}$ (C) $\dfrac{1}{6}$ (D) $\dfrac{5}{6}$ (E) $\dfrac{1}{2}$

Solution. Answer. (A)

Let (x, y) be a point in the square $OABC$ ($[0,1] \times [0,1]$) in the coordinate plane. We have that $y \in [0, 1]$.

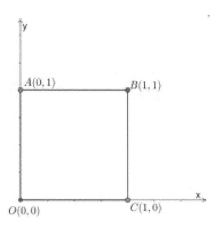

Thus, it follows that $y - 1.5 < 0$ and the given inequality is equivalent to the following inequality:
$$|x - 0.5| \leq y - 0.5,$$

or equivalently
$$0.5 - y \leq x - 0.5 \leq y - 0.5.$$

Hence, we obtain that $y \geq x$ and $y \geq 1 - x$. The solution of the last two inequalities is the shaded region ($\triangle ABD$).

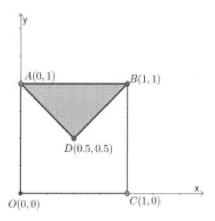

Therefore, the required probability is equal to $\dfrac{1}{4}$. □

Problem 4.167. *Given 22 circles, such that 21 of them have radius 1 (see the figure). Given also that any two circles that have a common point are pairwise tangent. What is the value of the circumference of the largest circle?*

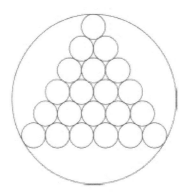

(A) $\dfrac{2\pi}{3}(10\sqrt{3}+3)$ (B) $\dfrac{20}{3}\pi$ (C) $\dfrac{20\sqrt{3}}{3}\pi$ (D) 13π (E) 20π

Solution. Answer. (A)
We have equilateral triangles with side lengths 2 (see the figure).

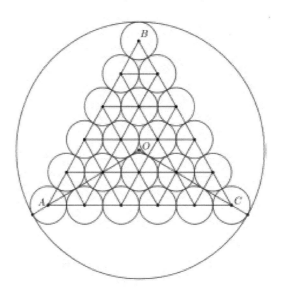

Therefore
$$AB = BC = AC = 10.$$

Let O be the center of the largest circle and R its radius. Then O is $(R-1)$ away from points A, B and C.
Thus, it follows that
$$R - 1 = \dfrac{10}{\sqrt{3}},$$
and
$$2\pi R = \dfrac{2\pi}{3}(10\sqrt{3}+3).$$

□

Problem 4.168. *How many seven-digit numbers contain 3 ones, 2 twos, 2 threes and no two neighboring digits are ones?*

(A) 90 (B) 120 (C) 60 (D) 100 (E) 150

Solution. Answer. (C)
The total number of seven-digit numbers having 3 ones, 2 twos, and 2 threes is
$$\frac{7!}{3!2!2!} = 210.$$
Among them, the total number of numbers containing 3 ones next to each other is
$$\frac{5!}{1!2!2!} = 30.$$
On the other hand, the total number of numbers such that only 2 of ones are next to each other is
$$4 \cdot 6 + 3 \cdot 6 + 3 \cdot 6 + 3 \cdot 6 + 3 \cdot 6 + 4 \cdot 6 = 120.$$
Therefore, the required total number is $210 - 30 - 120 = 60$. □

Problem 4.169. *A rectangular 2×10 grid is randomly covered by ten 1×2 rectangles (dominos). What is the probability that the two squares in the fifth column from the left will be covered by different dominos?*

(A) $\dfrac{64}{89}$ (B) $\dfrac{49}{89}$ (C) $\dfrac{25}{89}$ (D) $\dfrac{40}{89}$ (E) $\dfrac{1}{2}$

Solution. Answer. (B)
Let us denote by a_n the number of ways to cover $2 \times n$ rectangular grid by n dominos. We have that
$$a_1 = 1, a_2 = 2,$$
and
$$a_{n+2} = a_{n+1} + a_n, n = 1, 2, \ldots.$$
This is because there are only two ways to cover the left bottom square of $2 \times n$ rectangular grid.

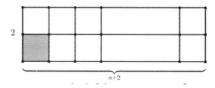

Note that if the 5th column from the left is covered by different dominos, then the following two cases are possible.

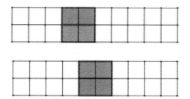

Thus, the required probability is:
$$p = \frac{a_3 \cdot a_5 + a_4 \cdot a_4}{a_{10}} = \frac{3 \cdot 8 + 5 \cdot 5}{89} = \frac{49}{89}.$$

□

Problem 4.170. *Let the vertices of triangle ABC lie on the surface of a sphere with a radius of 13 (see the figure). Given that $AB = 5$ and $m\angle ACB = 30°$. What is the value of the distance from the center of the sphere to the plane containing triangle ABC?*

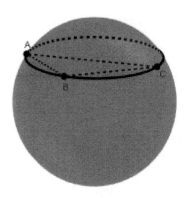

(A) 10   (B) 11   (C) 12   (D) 6   (E) 8

Solution. Answer. (C)

Let point O be the center of the sphere and point O_1 be the circumcenter of triange ABC (see the figure). Note that O_1 is not necessarily located inside of triangle ABC as it is shown in the figure.

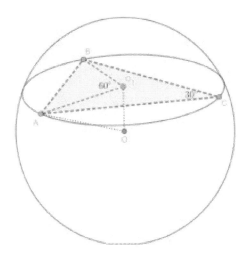

We have that
$$\angle AO_1B = 2\angle ACB = 60°,$$
therefore $\triangle ABO_1$ is an equilateral triangle. Thus, it follows that
$$AO_1 = AB = 5.$$

We have also $OO_1 \perp (ABC)$. Hence, we deduce that $\angle AO_1O = 90°$.
Therefore, from triangle AO_1O according to the Pythagorean theorem, we obtain that
$$OO_1 = \sqrt{AO^2 - AO_1^2} = 12.$$

□

Problem 4.171. *Let $f(x)$ be a monic polynomial function of degree five, such that*

$$f(2021) = -2, f(2022) = 1, f(2023) = 2, f(2024) = 1, f(2025) = -2.$$

What is the value of $f(2020)$?

(A) -5   (B) -12   (C) -127   (D) -210   (E) -2100

Solution. Answer. (C)
Let us consider the following function

$$h(x) = f(x) + (x - 2023)^2 - 2.$$

Note that $h(x)$ is a monic polynomial function of degree five too and

$$h(2021) = h(2022) = h(2023) = h(2024) = h(2025) = 0.$$

Thus, it follows that 2021, 2022, 2023, 2024, 2025 are roots of $h(x)$. We have that $h(x)$ is a monic polynomial function of degree five, therefore besides these five roots it does not have any other root. Hence, according to the linear factorization theorem we obtain that

$$h(x) = (x - 2021)(x - 2022)(x - 2023)(x - 2024)(x - 2025).$$

We deduce that

$$f(x) = (x - 2021)(x - 2022)(x - 2023)(x - 2024)(x - 2025) - (x - 2023)^2 + 2.$$

Thus, it follows that

$$f(2020) = (-1)(-2)(-3)(-4)(-5) - (-3)^2 + 2 = -120 - 9 + 2 = -127.$$

Alternative solution. According to Newton's divided difference formula, we have that

$$f(x) = (x-2021)(x-2022)(x-2023)(x-2024)(x-2025) + a(x-2021)(x-2022)(x-2023)(x-2024)+$$

$$+b(x-2021)(x-2022)(x-2023) + c(x-2021)(x-2022) + d(x-2021) + e,$$

where a, b, c, d, e are some numbers.
Note that

$$\begin{cases} e = f(2021) = -2, \\ e + d = f(2022) = 1, \\ e + 2d + 2c = f(2023) = 2, \\ e + 3d + 6c + 6b = f(2024) = 1, \\ e + 4d + 12c + 24b + 24a = f(2025) = -2. \end{cases}$$

Solving this system of equations, we deduce that

$$e = -2, d = 3, c = -1, b = 0, a = 0.$$

Thus, it follows that

$$f(x) = (x-2021)(x-2022)(x-2023)(x-2024)(x-2025) - (x-2021)(x-2022) + 3(x-2021) - 2.$$

We obtain that

$$f(2020) = -120 - 2 - 3 - 2 = -127.$$

□

Problem 4.172. Let (a_n) be a number sequence defined as follows: $a_1 = a_2 = 1, a_3 = \dfrac{1}{2}$ and $a_{n+1} = \dfrac{a_n^3 a_{n-2}}{a_{n-1}^3 + a_{n-2}a_{n-1}a_n}$, where $n = 3, 4, \ldots$ What is the value of

$$S_{10} = \frac{a_1}{a_2} + 2\frac{a_2}{a_3} + 3\frac{a_3}{a_4} + \ldots + 10\frac{a_{10}}{a_{11}}.$$

(A) $11!$ (B) $11! - 1$ (C) $10! + 3$ (D) $10! + 11$ (E) $10 \cdot 11$

Solution. Answer. (B)

The given condition can be rewritten in the following way:

$$\frac{a_n^2}{a_{n-1}a_{n+1}} = \frac{a_{n-1}^2}{a_{n-2}a_n} + 1,$$

where $n = 2, 3, \ldots$ Let us consider the following sequence

$$b_n = \frac{a_{n+1}^2}{a_n a_{n+2}},$$

where $n = 1, 2, 3, \ldots$ We have that

$$b_1 = \frac{a_2^2}{a_1 a_3} = 2,$$
$$b_{n-1} = b_{n-2} + 1,$$

where $n = 3, 4, \ldots$ Thus, it follows that $b_n = n + 1$, where $n = 1, 2, \ldots$ We have that

$$\frac{a_2^2}{a_1 a_3} = 2, \frac{a_3^2}{a_2 a_4} = 3, \ldots, \frac{a_{n+1}^2}{a_n a_{n+2}} = n + 1.$$

Multiplying all the equations we obtain that

$$\frac{a_{n+1}}{a_{n+2}} = (n+1)!.$$

On the other hand, we have that

$$\frac{a_2}{a_3} = 2\frac{a_1}{a_2}, \frac{a_3}{a_4} = 3\frac{a_2}{a_3}, \ldots, \frac{a_{n+1}}{a_{n+2}} = (n+1)\frac{a_n}{a_{n+1}}.$$

Summing up the last equations we deduce that

$$S_n = \frac{a_1}{a_2} + 2\frac{a_2}{a_3} + 3\frac{a_3}{a_4} + \ldots + n\frac{a_n}{a_{n+1}} = (n+1)! - 1.$$

Thus, it follows that

$$S_{10} = 11! - 1.$$

□

Problem 4.173. Let a, b, c be integers, such that $x^2 + 3x + 2 \leq ax^2 + bx + c \leq 2x^2 + 12x + 25$ double inequality holds true for all values of x. What is the value of the sum of all possible values of c?

(A) 60   (B) 61   (C) 80   (D) 81   (E) 90

Solution. Answer. (D)
Note that
$$mx^2 + nx + p \geq 0$$
inequality holds true for all values of x, then either $m = 0, n = 0, p \geq 0$ or $m > 0, n^2 - 4mp \leq 0$. Then, we deduce that either $a = 1$ or $a = 2$. Let us do the following casework:
Case 1. $a = 1, b = 3$ and $c \geq 2$,
$$x^2 + 9x + 25 - c \geq 0,$$
condition reveals that
$$81 - 4(25 - c) \leq 0,$$
or $c \in \{2, 3, 4\}$.
Case 2. $a = 2, b = 12$ and $c \leq 25$,
$$x^2 + 9x + c - 2 \geq 0$$
condition reveals that
$$81 - 4(c - 2) \leq 0$$
or $c \in \{23, 24, 25\}$. Therefore, the sum of all possible values of c is
$$2 + 3 + 4 + 23 + 24 + 25 = 81.$$

□

Problem 4.174. *A league of soccer teams participate in a tournament. Any two teams play each other exactly once. By the end of the tournament, exactly n games end in a tie and the total points gained in the tournament is 2019. What is the value of the sum of all possible values of n? (Note: A winning team gets 3 points, a losing team gets 0 points, while a tie is 1 point.)*

(A) 2019   (B) 2050   (C) 2150   (D) 4038   (E) 4080

Solution. Answer. (E)
Let m be the number of teams participating in the tournament. Thus, it follows that during the tournament in total $\dfrac{m(m-1)}{2}$ games were played. Given that
$$2n + 3\left(\dfrac{m(m-1)}{2} - n\right) = 2019,$$
or equivalently
$$n = \dfrac{3m(m-1)}{2} - 2019.$$
Note that $0 \leq n \leq \dfrac{m(m-1)}{2}$. Therefore
$$1346 \leq m(m-1) \leq 2019.$$
From the last double-inequality we obtain that
$$m \in \{38, 39, 40, 41, 42, 43, 44, 45\},$$
and
$$n \in \{\dfrac{3(38^2 - 38)}{2} - 2019, 3(39^2 - 39)2 - 2019, ..., 3(45^2 - 45)2 - 2019.\}$$

Now, we need to calculate the following:

$$\frac{3}{2}(38^2+39^2+\ldots+45^2-38-39-\ldots-45)-8\cdot 2019 = \frac{3}{2}\left(\frac{45\cdot 46\cdot 91}{6} - \frac{37\cdot 38\cdot 75}{6} - \frac{38+45}{2}\cdot 8\right) - 8\cdot 2019 = 4080.$$

$\square$

Problem 4.175. *Let n be a positive integers such that $\dfrac{(n!)^2}{(n+3)!}$ is also a positive integer. What is the smallest possible number of divisors of $(n+1)(n+2)(n+3)$?*

(A) 24 (B) 30 (C) 36 (D) 12 (E) 20

Solution. Answer. (B)
Given that $[(n+1)(n+2)(n+3)] \mid n!$, as we have that

$$\frac{(n!)^2}{(n+3)!} = \frac{n!}{(n+1)(n+2)(n+3)}.$$

Therefore, each of the numbers $n+1, n+2$, and $n+3$ is not prime. Let us do a casework:
Case 1. n is even. Then any two of the numbers $n+1, n+2, n+3$ are relatively prime. Let

$$n+1 = ab, n+2 = cd, n+3 = ef.$$

Moreover
$$1 < a \le b, 1 < c \le d, 1 < e \le f$$

and only one of the consecutive numbers $n+1, n+2, n+3$ can be a perfect square. Hence, if we replace each of the numbers a, b, c, d, e and f with their largest prime factor, we get that the number of divisors of the number $abcdef$ is not less than $2\cdot 2\cdot 2\cdot 2\cdot 3 = 48$.
Case 2. n is odd. Then one of the numbers $n+1$ and $n+3$ is even, while the other one is multiple of 4, larger than 4. Thus, it follows that

$$(n+1)(n+2)(n+3) = 8uvcd,$$

where each of the numbers u, v, c and d is larger than 1 and

$$(u,v) = 1, (u,c) = 1, (u,d) = 1, (v,c) = 1, (v,d) = 1,$$

c and d are odd numbers ($n+2 = cd$). Therefore, if we replace each of the numbers u, v, c and d with their largest prime factor, we obtain that the number of divisors of $8uvcd$ is less than $5\cdot 2\cdot 3 = 30$.
When $n = 7$, we have that
$$\frac{(n!)^2}{(n+3)!} = 7,$$

and
$$8\cdot 9\cdot 10 = 2^4\cdot 3^2\cdot 5$$

has $(4+1)(2+1)(1+1) = 30$ divisors. $\square$

4.8 Solutions of AMC 10 type practice test 8

Problem 4.176. *Some of the students in the class are students in a math circle as well. Most of the students in the class participated in the AMC 10 test. It appeared that 7⁄8 of the math circle students and 5⁄6 of students not attending the math circle participated in the AMC 10. What part of the class participated in the AMC 10, if the total number of students in the class is not more than 19?*

(A) $\dfrac{1}{2}$ (B) $\dfrac{2}{3}$ (C) $\dfrac{6}{7}$ (D) $\dfrac{3}{4}$ (E) $\dfrac{8}{9}$

Solution. Answer. (C)
Given that the number of math circle students in the class is a multiple of 8, while the number of students not attending the math circle is a multiple of 6. As the total number of the students in the class is not more than 19, then there are 8 students attending the math circle and 6 students not attending the math circle. Thus, the required ratio is
$$\frac{7+5}{8+6} = \frac{6}{7}.$$

□

Problem 4.177. *There are 360 books in the library. 20% of the books that are about mathematics and 50% of the books that are not about mathematics are hard cover. Given that 0.3 part of the books in the library are hard cover, how many books about mathematics are in the library?*

(A) 240   (B) 120   (C) 180   (D) 270   (E) 300

Solution. Answer. (A)
Let the number of math books in the library be x, then the number of non-math books is $360 - x$. Therefore, the number of math books with hard cover is $\dfrac{x \cdot 20}{100}$, while the number of non-math books is
$$\frac{(360-x) \cdot 50}{100}.$$
Given that
$$\frac{x}{5} + \frac{360-x}{2} = 360 \cdot 0.3.$$
Thus, it follows that $1800 - 3x = 1080$. Therefore $x = 240$.

□

Problem 4.178. *Given that the mean of $30, 31, ..., 29 + 2n$ is equal to the mean of $100, 101, ..., 99 + n$. What is the value of the sum of the digits of n?*

(A) 10   (B) 5   (C) 8   (D) 7   (E) 6

Solution. Answer. (B)
Given that
$$30 + 29 + 2n = 100 + 99 + n.$$
Thus, it follows that $n = 140$.
Therefore, the sum of the digits of n is $1 + 4 + 0 = 5$.

□

Problem 4.179. *Two boxes contain an equal number of pens. Given that all pens are either white or black. The ratio of white to black pens in the first box is 7 : 3 and 3 : 2 in the second box. If the total number of black pens is 35, how many white pens are in the first box?*

(A) 20 (B) 25 (C) 30 (D) 35 (E) 40

Solution. Answer. (D)
Let the number of pens in the first box be x. The amount of black pens in the first and second box is $\frac{3}{10}x$ and $\frac{2}{5}x$, respectively.
We have that
$$\frac{3}{10}x + \frac{2}{5}x = 35.$$
Thus, it follows that $x = 50$.
Hence, we obtain that
$$\frac{7}{10}x = 35.$$
$\square$

Problem 4.180. *For how many of the following quadrilaterals does there never exist a point (on the plane of that quadrilateral) that is equidistant from all four sides of that quadrilateral?*
- *A rhombus that is not a square.*
- *A parallelogram that is not a rectangle or a rhombus.*
- *An isosceles trapezoid that is not a parallelogram.*
- *Not isosceles trapezoid.*
- *quadrilateral that is neither a parallelogram nor a trapezoid.*

(A) 1 (B) 2 (C) 3 (D) 4 (E) 5

Solution. Answer. (A)
The point that is equidistant from all four sides is the center of the inscribed circle.
The main condition for this is $a + c = b + d$, where a, b, c, d are consecutive sides of the quadrilateral. Note that this condition holds true for any rhombus.
It may hold true or may not hold true for other shapes except a parallelogram that is not a rhombus. In the parallelogram (that is not a rhombus), the sum of two longer sides is greater than the sum of two shorter ones. Therefore, only in this case this point never exists. Hnece, the answer is 1. $\square$

Problem 4.181. *Given that $n!, (n+1)!$ and $n!(n+19)$ form an arithmetic sequence where n is a positive integer. What is the product of all digits of n?*

(A) 18 (B) 24 (C) 30 (D) 3 (E) 8

Solution. Answer. (E)
We have that
$$n! + n!(n+19) = 2(n+1)!.$$
Thus, it follows that
$$1 + (n+19) = 2n + 2.$$
Therefore $n = 18$ and the product of all digits of n is equal to 8. $\square$

Problem 4.182. *Suppose that lines $ax+by=c$, $(a+2)x+(b-3)y=c+9$ and $(a+1)x+(b-1)y=c+2$ are concurrent at point M. What is the value of the sum of the coordinates of M?*

(A) 0 (B) 1 (C) -1 (D) -2 (E) -8

Solution. Answer. (E)
Let $M(x_0, y_0)$. Given that
$$\begin{cases} ax_0 + by_0 = c, \\ (a+2)x_0 + (b-3)y_0 = c+9, \\ (a+1)x_0 + (b-1)y_0 = c+2. \end{cases}$$

Thus, it follows that
$$\begin{cases} 2x_0 - 3y_0 = 9, \\ x_0 - y_0 = 2. \end{cases}$$

Hence, we obtain that $y_0 = -5$ and $x_0 = -3$. Therefore
$$x_0 + y_0 = -8.$$

$\square$

Problem 4.183. *Positive integers m and n are such that both the sum and the difference of $mn+m+n-53$ and $9n-6m+3$ are prime numbers. What is the value of the sum of the digits of the product mn?*

(A) 7 (B) 9 (C) 10 (D) 13 (E) 15

Solution. Answer. (B)
Given that
$$mn + 7m - 8n - 56 = (m-8)(n+7),$$
and
$$mn - 5m + 10n - 50 = (m+10)(n-5)$$
are prime numbers. Note that
$$n + 7 \geq 8 > 1,$$
and
$$m + 10 \geq 11 > 1.$$
Therefore $m - 8 = 1$ and $n - 5 = 1$.
Thus, it follows that $m = 9, n = 6$ and
$$(m-8)(n+7) = 13,$$
$$(m+10)(n-5) = 19.$$
Hence, we obtain that $mn = 54$.

$\square$

Problem 4.184. *Given 1×2 and 3×6 rectangles (see the figure). What is the value of the area of $\triangle ABC$?*

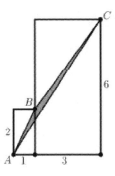

(A) 2   (B) 3   (C) 1   (D) 1.5   (E) 0.5

Solution. Answer. (C)

Note that $\triangle ABD \sim \triangle DEC$, as $\dfrac{1}{3} = \dfrac{2}{6}$ and
$$\angle ADB = \angle DEC = 90°.$$

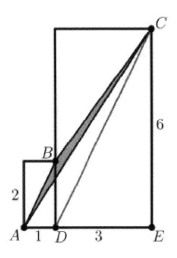

Thus, it follows that $\angle BAD = \angle CDE$. Therefore $AB \parallel CD$. Hence, we obtain that
$$[ABC] = [ABD] = \dfrac{1 \cdot 2}{2} = 1.$$

Alternative solution. More primitive solution (but not so beautiful) is also possible. Let us subtract the areas of three triangles from the total area of the figure, then we obtain that
$$(2 \cdot 1 + 3 \cdot 6) - \dfrac{4 \cdot 6}{2} - \dfrac{1 \cdot 2}{2} - \dfrac{3 \cdot 4}{2} = 20 - 12 - 1 - 6 = 1.$$

□

Problem 4.185. *Let function f is defined by $f(x) = \lfloor 3x \rfloor - \lfloor 2x \rfloor - \lfloor x \rfloor$ for all real numbers x, where $\lfloor r \rfloor$ denotes the greatest integer that is less than or equal to the real number r. What is the range of f?*

(A) $\{0\}$ (B) $\{1\}$ (C) $\{0,1\}$ (D) $\{0,1,2\}$ (E) $\{0,2\}$

Solution. Answer. (C)
Let $x = \lfloor x \rfloor + \{x\}$. Note that
$$\lfloor a + b \rfloor = \lfloor a \rfloor + b$$
for any real number a and any integer b.
Thus, it follows that
$$f(x) = \lfloor 3x \rfloor - \lfloor 2x \rfloor - \lfloor x \rfloor = \lfloor 3(\lfloor x \rfloor + \{x\}) \rfloor - \lfloor 2(\lfloor x \rfloor + \{x\}) \rfloor - \lfloor x \rfloor =$$
$$= 3\lfloor x \rfloor + \lfloor 3\{x\} \rfloor - 2\lfloor x \rfloor - \lfloor 2\{x\} \rfloor - \lfloor x \rfloor = \lfloor 3\{x\} \rfloor - \lfloor 2\{x\} \rfloor.$$

Hence, we obtain that
$$f(x) = \begin{cases} 0, if\ 0 \leq \{x\} < \dfrac{1}{3}, \\ 1, if\ \dfrac{1}{3} \leq \{x\} < \dfrac{1}{2}, \\ 0, if\ \dfrac{1}{2} \leq \{x\} < \dfrac{2}{3}, \\ 1, if\ \dfrac{2}{3} \leq \{x\} < 1. \end{cases}$$

Therefore, we deduce that $E(f) = \{0, 1\}$. $\square$

Problem 4.186. *Let the lengths of two of the sides of $\triangle ABC$ be 6, 8 and its area be 24. What is the value of the perimeter of $\triangle ABC$?*

(A) 20   (B) 24   (C) 22   (D) 27   (E) 26

Solution. Answer. (B)
Let $AB = 6$ and $BC = 8$. We have that
$$[ABC] = \frac{1}{2} AB \cdot h_c,$$
or
$$h_c = 8 = BC.$$

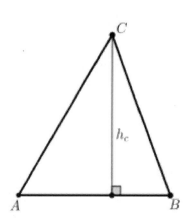

Thus, it follows that $\angle ABC = 90°$. According to Pythagorean theorem, we have that
$$AC^2 = 6^2 + 8^2 = 10^2.$$
Hence, we deduce that $AC = 10$.
Therefore, the perimeter of $\triangle ABC$ is $6 + 8 + 10 = 24$. $\square$

Problem 4.187. *For how many positive integers n, both fractions $\dfrac{20n+19}{n+817}$ and $\dfrac{19n+61}{n+817}$ are positive integers too?*

(A) 0 (B) 1 (C) 3 (D) 5 (E) 11

Solution. Answer. (B)
Given that
$$\frac{20n+19}{n+817} - \frac{19n+61}{n+817} = \frac{n-42}{n+817}.$$
is an integer. Thus, it follows that $n = 42$. Hence, we deduce that
$$\frac{20n+19}{n+817} = 1,$$
and
$$\frac{19n+61}{n+817} = 1.$$
Therefore, there is only one such value of n. $\square$

Problem 4.188. *Suppose you write the set $\{4000, 4001, 4002, ...5000\}$ in base 10. Then you recorded this set in base 5 and found the number in the new set whose digits have the least sum. What is the value of the sum of the digits of this number?*

(A) 7 (B) 6 (C) 5 (D) 4 (E) 3

Solution. Answer. (E)
We have that
$$4000 = 2^5 \cdot 5^3 = (1 \cdot 25 + 1 \cdot 5 + 2) \cdot 5^3,$$
and
$$5000 = 2^3 \cdot 5^4 = (1 \cdot 5 + 3) \cdot 5^4,$$
then $4000 = 112000_5$ and $5000 = 130000_5$.
Therefore, the number in the new set of base 5 system with the smallest sum of the digits is 120000_5 and the sum of the digits is 3. $\square$

Problem 4.189. *Given that $19! = 121,645,100,408,ab2,000$, what is the largest prime factor of the number $\overline{ab2}$?*

(A) 11 (B) 7 (C) 13 (D) 17 (E) 5

Solution. Answer. (C)
We have that $9 \mid 19!$ and $11 \mid 19!$, therefore $9 \mid (34 + a + b)$ and $11 \mid (9 + a - (25 + b))$. Thus, it follows that $9 \mid (7 + a + b)$ and $11 \mid (a - b - 16)$.
Hence, we obtain that $a + b \in \{2, 11\}$ and $a - b \in \{-6, 5\}$.
We deduce that $a + b = 11$ and $a - b = 5$.
Thus, it follows that $a = 8, b = 3$. Hence, we obtain that
$$\overline{ab2} = 832 = 2^6 \cdot 13.$$
Therefore, the largest prime factor of 832 is 13. $\square$

Problem 4.190. *Given that the area of the triangle with sides a, b, c and the area of the triangle with sides a, b, d, where $d \neq c$, are equal. Given also that $c^2 + d^2 = 2020$. What is the value of $a^2 + b^2$?*

(A) 505   (B) 1010   (C) 101   (D) 400   (E) 640

Solution. Answer. (B)
We have that the opposite angles of sides c and d are not equal, however their sines are equal and this means the sum of these angles is equal to $180°$. Therefore, we can consider the following "composite" triangle, where the median is b.

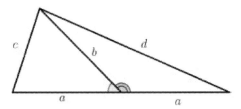

According to the formula for a median, we have that
$$4a^2 + 4b^2 = 2c^2 + 2d^2.$$

Thus, it follows that
$$a^2 + b^2 = \frac{c^2 + d^2}{2} = 1010.$$

Alternative solution. We have that the opposite angles of sides c and d are not equal, however their sines are equal. Therefore $\cos\theta = -\cos\gamma$. From the law of cosines, it follows that
$$c^2 = a^2 + b^2 - 2ab\cos\gamma, d^2 = a^2 + b2 + 2ab\cos\gamma.$$

Summing up the last two equations, we obtain that
$$c^2 + d^2 = 2(a^2 + b^2).$$

Thus, it follows that
$$a^2 + b^2 = \frac{c^2 + d^2}{2} = 1010.$$

□

Problem 4.191. *Let points D and E lie on sides AC and BC of triangle ABC, respectively. Given that $\angle BAC = 30°, \angle ACB = 15°, AB = BD$ and $DE = EC$. What is the ratio $AD : CD$?*

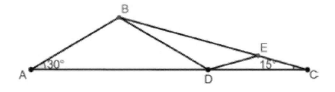

(A) $\sqrt{3} : 1$ (B) $1 : \sqrt{3}$ (C) $1 : 2$ (D) $2 : 1$ (E) $1 : 3$

Solution. Answer. (A)
We have that
$$\angle EDC = \angle ACB = 15°,$$
thus $\angle BED = 30°$. Note that
$$\angle BDA = \angle BAD = 30°,$$
and $\angle BDE = 135°$. Hence, we obtain that
$$\angle DBE = 180° - 135° - 30° = 15° = \angle DCB.$$

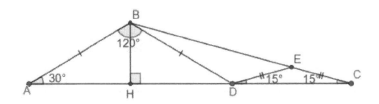

Therefore $BD = DC$. Thus, it follows that
$$\frac{AD}{CD} = \frac{AD}{BD} = \frac{2 \cdot HD}{BD} = 2\cos 30° = \sqrt{3}.$$

□

Problem 4.192. *Let A denote the smallest positive integer that is divisible by 36 and whose base 10 representation consists of only 4's and 9's, with at least one of each. What is the value of the remainder of A after a division by 11?*

(A) 1   (B) 3   (C) 5   (D) 8   (E) 10

Solution. Answer. (C)
This number A is divisible by 36, therefore $9 \mid A, 4 \mid A$.
From the condition $4 \mid A$ it follows that the last two digits have to be divisible by 4. That means the last two digits of this integer are 4 and 4.
From the condition $9 \mid A$ it follows that the sum of all the digits is divisible by 9.
Hence, we obtain that
$$9 \mid (4 + 4 + 4 \cdot n + 9 \cdot (8 - n)).$$
We deduce that $9 \mid (1 + 5n)$. Thus, it follows that $n = 7$.
We have $(2 + 7) = 9$ fours and one nine.
The smallest integer that satisfies all these conditions is 4444444944.
The remainder after the division by 11 is 5, as
$$4444444944 (4 + 4 + 4 + 9 + 4) - (4 \cdot 5) 5 (mod 11).$$

□

Problem 4.193. *A hotel with infinite number of rooms has room numbers labeled with the positive integers 1, 2, 3,... A guest is assigned to room i with probability $\frac{1}{2^i}$. What is the probability that the positive difference of the room numbers assigned to two random guests is 2?*

(A) $\frac{1}{3}$ (B) $\frac{2}{3}$ (C) $\frac{1}{12}$ (D) $\frac{1}{6}$ (E) $\frac{1}{4}$

Solution. Answer. (D)
Note that the probability of the event that the first guest is assigned the room i (where $i \geq 3$) and the second guest is assigned either the room $i-2$ or $i+2$ equals to

$$\frac{1}{2^i} \cdot \frac{1}{2^{i-2}} + \frac{1}{2^i} \cdot \frac{1}{2^{i+2}}.$$

Therefore, the required probability is

$$\frac{1}{2} \cdot \frac{1}{2^3} + \frac{1}{2^2} \cdot \frac{1}{2^4} + \left(\frac{1}{2^3} \cdot \frac{1}{2} + \frac{1}{2^3} \cdot \frac{1}{2^5}\right) + \left(\frac{1}{2^4} \cdot \frac{1}{2^2} + \frac{1}{2^4} \cdot \frac{1}{2^6}\right) + ... =$$

$$= \frac{2}{2 \cdot 2^3} + \frac{2}{2^2 \cdot 2^4} + \frac{2}{2^3 \cdot 2^5} + ... = \frac{1}{2^3} + \frac{1}{2^5} + ... = \frac{\frac{1}{8}}{1 - \frac{1}{4}} = \frac{1}{6}.$$

□

Problem 4.194. *Given that point $M(x_0, y_0)$ and circle σ lie on the same plane. Secants MB and MD are drawn such that the rays BM, DM meet σ at points A, C, respectively. Given that $AB = CD$, $A(2, 4), B(3, 6)$ and $C(0, 2)$. What is the value of $4x_0 + y_0$?*

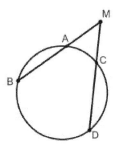

(A) 2 (B) 3 (C) 8 (D) 0 (E) 5

Solution. Answer. (C)
Let O be the center of circle σ. From the condition $AB = CD$ it follows that $\triangle OAB$ is congruent to $\triangle OCD$. Hence, we obtain that O is equidistant from the sides of $\angle BMD$. Therefore $MA = MC$. Note that points A and B lie on the line $y = 2x$, so M also lies on it. Thus, it follows that

$$\begin{cases} y_0 = 2x_0, \\ \sqrt{(x_0 - 2)^2 + (y_0 - 4)^2} = \sqrt{(x_0 - 0)^2 + (y_0 - 2)^2}. \end{cases}$$

Hence, we deduce that

$$\begin{cases} y_0 = 2x_0, \\ x_0 + y_0 = 4. \end{cases}$$

Therefore, we get that

$$x_0 = \frac{4}{3}, y_0 = \frac{8}{3}.$$

Thus, it follows that
$$4x_0 + y_0 = 4 \cdot \frac{4}{3} + \frac{8}{3} = 8.$$
$\square$

Problem 4.195. *Given that 10% of the numbers in the set $\{1, 2, ..., n\}$ is divisible by 9. What is the total number of all possible values of n?*

(A) 2   (B) 9   (C) 13   (D) ∞ (E) 8

Solution. Answer. (E)
We need to solve the following equation
$$\lfloor \frac{n}{9} \rfloor = \frac{n}{10}.$$
Thus, it follows that $n = 10k$, where k is a positive integer. Therefore, plugging in $n = 10k$ in the first equation, we obtain that
$$\lfloor k + \frac{k}{9} \rfloor = k.$$
Hence
$$k + \lfloor \frac{k}{9} \rfloor = k.$$
We deduce that
$$\lfloor \frac{k}{9} \rfloor = 0.$$
Thus, it follows that $1 \leq k \leq 8$. Therefore, there are 8 different possible values of n. $\square$

Problem 4.196. *Let a be the least possible value of x, such that the values of the functions $y = \frac{5}{6}x + \frac{1}{3}$ and $y = \frac{25}{16}x - \frac{1}{2}$ are positive integers. What is the value of the sum of the digits of a^3?*

(A) 1   (B) 2   (C) 3   (D) 6   (E) 8

Solution. Answer. (E)
Let
$$\frac{5}{6}a + \frac{1}{3} = m,$$
and
$$\frac{25}{16}a - \frac{1}{2} = k,$$
where m and k are positive integers.
We need to find the least possible value of m. We have that
$$a = \frac{6m - 2}{5},$$
and
$$a = \frac{16}{25}k + \frac{8}{25}.$$
Hence, we deduce that
$$\frac{6m - 2}{5} = \frac{16}{25}k + \frac{8}{25},$$
or equivalently $30m - 10 = 16k + 8$. Therefore $15m - 9 = 8k$. Thus, it follows that the solutions are
$$m = 7 + 8q, k = 12 + 15q,$$
where q is a non-negative integer.
Hence, the least possible value of m is 7. We obtain that $a = 8$ and $a^3 = 512$. Thus, the sum of the digits of a^3 is equal to 8. $\square$

Problem 4.197. Let ABC be a triangle, such that $AC = 3, BC = 5, AB = 7$ and its each side is the diameter of a semicircle (see the figure). The areas of the grey figures are labeled S_1, S_2 and S_3. What is the value of $S_1 + S_3 - S_2$?

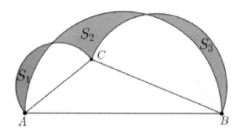

(A) $\dfrac{15}{8}(2\sqrt{3} - \pi)$ (B) $\dfrac{15}{4}(2\sqrt{3} - \pi)$ (C) $\dfrac{\pi}{16} - \dfrac{\sqrt{3}}{18}$ (D) 4 (E) $\pi + 2$

Solution. Answer. (A)
Note that
$$S_1 + S_3 + \frac{\pi \cdot 3.5^2}{2} = \frac{\pi \cdot 1.5^2}{2} + \frac{\pi \cdot 2.5^2}{2} + S_2 + [ABC].$$
According to Heron's formula, we have that
$$[ABC] = \sqrt{\frac{15}{2} \cdot \frac{1}{2} \cdot \frac{9}{2} \cdot \frac{5}{2}} = \frac{15\sqrt{3}}{4}.$$
Thus, it follows that
$$S_1 + S_3 - S_2 = -\frac{\pi}{2}(3.5^2 - 1.5^2 - 2.5^2) + \frac{15\sqrt{3}}{4} = \frac{15}{8}(2\sqrt{3} - \pi).$$

□

Problem 4.198. *How many numbers $\overline{a_1 a_2 ... a_{10}}$ with nonzero digits exist, such that each of the numbers $\overline{a_1 a_2 a_3}, \overline{a_2 a_3 a_4}, ..., \overline{a_8 a_9 a_{10}}$ is not divisible by 3?*

(A) $2^8 \cdot 3^{10}$ (B) $2^8 \cdot 3^{12}$ (C) $2^{10} \cdot 3^{10}$ (D) $2^{12} \cdot 3^{12}$ (E) $2^{12} \cdot 3^8$

Solution. Answer. (B)
Note that both a_1 and a_2 can be chosen in 9 ways, while each of the numbers $a_3, ..., a_{10}$ in 6 ways. Therefore, the required quantity of 10-digit numbers is
$$9 \cdot 9 \cdot 6 \cdot ... \cdot 6 = 3^4 \cdot 6^8 = 3^{12} \cdot 2^8.$$

□

Problem 4.199. *A random four-digit number is chosen. What is the probability that the digits of the chosen number are four consecutive numbers?*

(A) $\dfrac{1}{30}$ (B) $\dfrac{9}{500}$ (C) $\dfrac{7}{300}$ (D) $\dfrac{11}{450}$ (E) $\dfrac{7}{80}$

Solution. Answer. (B)
Let us find the quantity of four-digit numbers with digits that are consecutive numbers.
The quantity of four-digit numbers with digits 0, 1, 2, 3 is $3 \cdot 3 \cdot 2 \cdot 1 = 18$.
The quantity of four-digit numbers with digits 1, 2, 3, 4 is $4 \cdot 3 \cdot 2 \cdot 1 = 24$.
In a similar way, we can find that the quantity of four-digit numbers with digits 6, 7, 8, 9 is $4 \cdot 3 \cdot 2 \cdot 1 = 24$.

Thus, the total quantity of four-digit numbers with the above mentioned property is $6 \cdot 24 + 18 = 162$.
In total, there are $9 \cdot 10 \cdot 10 \cdot 10 = 9000$ four digit numbers.
Therefore, the required probability is
$$\frac{162}{9000} = \frac{9}{500}.$$

□

Problem 4.200. *Let sequence* (x_n) *be defined as follows:* $x_1 = 1$ *and* $x_{n+1} = x_n + \frac{1}{x_n}$. *What is the value of the sum of the digits of the smallest possible* n, *such that* $x_n > 8$?

(A) 5   (B) 12   (C) 16   (D) 21   (E) 24

Solution. Answer. (A)
We have that
$$x_{i+1}^2 = x_i^2 + 2 + \frac{1}{x_i^2}.$$

Summing up all equations (for $i = 1, 2, ..., n$) we obtain that
$$x_{n+1}^2 = 2n + 2 + \frac{1}{x_2^2} + ... + \frac{1}{x_n^2}.$$

Thus, it follows that $x_{n+1}^2 > 2n + 2$ and $x_{32} > \sqrt{2 \cdot 31 + 2} = 8$.
On the other hand, we have that
$$x_{31}^2 = 62 + \frac{1}{x_2^2} + \frac{1}{x_3^2} + ... + \frac{1}{x_{30}^2} < 62 + \frac{1}{4} + \frac{1}{6} + ... + \frac{1}{60} <$$
$$< 62 + \frac{1}{2}\left(\left(\frac{1}{2} + \frac{1}{3}\right) + \left(\frac{1}{4} + ... + \frac{1}{7}\right) + \left(\frac{1}{8} + ... + \frac{1}{15}\right) + \left(\frac{1}{16} + ... + \frac{1}{31}\right)\right) <$$
$$< 62 + \frac{1}{2}\left(2 \cdot \frac{1}{2} + 4 \cdot \frac{1}{4} + 8 \cdot \frac{1}{8} + 16 \cdot \frac{1}{16}\right) = 64.$$

Hence, we deduce that $x_{31} < 8$.
Therefore $n = 32$ and the sum of the digits is $3 + 2 = 5$.

□

4.9 Solutions of AMC 10 type practice test 9

Problem 4.201. *What is the value of the expression* $(((1^{-1} - 2^{-1})^{-2} - 3)^{-2} - 4)^{-2}$?

(A) 9 (B) 3 (C) 1 (D) $\frac{1}{3}$ (E) $\frac{1}{9}$

Solution. Answer. (E)
Note that
$$(((1^{-1} - 2^{-1})^{-2} - 3)^{-2} - 4)^{-2} = \left(\left(1 - \frac{1}{2}\right)^{-2} - 4\right)^{-2} = ((4-3)^{-2} - 4)^{-2} = (1-4)^{-2} = \frac{1}{9}.$$
□

Problem 4.202. *On the entrance of the shop is written* 20% + 20% *discount. It means, that the price is decreased by* 20% *and then the new price is again decreased by* 20%. *By what percent can the* 20% + 20% *discount be replaced?*

(A) 40 (B) 38 (C) 36 (D) 35 (E) 34

Solution. Answer. (C)
Let the initial price of the product be 100 units. After decreasing by 20% the new price is 80 units. Once again after 20% discount, the new price is $\frac{80 \times 80}{100} = 64$ units.
Therefore, 20% + 20% discount can be replaced by 36% discount. □

Problem 4.203. *For how many different values of variable x the expression* $\dfrac{1}{2 - \dfrac{3}{4 - \dfrac{1}{x^2}}}$ *is not defined?*

(A) 1 (B) 2 (C) 3 (D) 4 (E) 5

Solution. Answer. (E)
Note that the given expression is not defined when any of the denominators is equal to 0. Therefore, it is not defined only and only if
$$x^2 = 0,$$
or
$$4 - \frac{1}{x^2} = 0,$$
or
$$2 - \frac{3}{4 - \dfrac{1}{x^2}} = 0.$$

We obtain that either $x = 0$ or $x = -\frac{1}{2}$ or $x = \frac{1}{2}$ or $x = -\frac{\sqrt{10}}{5}$ or $x = \frac{\sqrt{10}}{5}$.
Therefore, the given expression is not defined for 5 different values of variable x. □

Problem 4.204. *Given that one of the numbers 1, 2, ..., 9 is equal to the arithmetic mean of the other eight numbers. What is the value of the sum of the digits of the sum of those eight number?*

(A) 2   (B) 3   (C) 4   (D) 6   (E) 11

Solution. Answer. (C)
Let n be the arithmetic mean of those 8 numbers, therefore
$$n = \frac{1+2+...+9-n}{8}.$$
Thus, it follows that
$$9n = 45.$$
Hence, we obtain that
$$n = 5.$$
Therefore, the sum of those eight numbers is $1+2+...+9-5 = 40$. Thus, the sum of the digits of the sum of those eight numbers is equal to 4, as $4+0 = 4$. $\square$

Problem 4.205. *Let n be a positive integer. What is the total number of all the elements of the set $\left\{1, \frac{1}{2}, \frac{1}{3}, ..., \frac{1}{n}, ...\right\}$, such that each of them is a solution of the inequality $100x^2 - 25x + 1 < 0$?*

(A) 13   (B) 14   (C) 15   (D) 16   (E) 19

Solution. Answer. (B)
Note that $\frac{1}{20}$ and $\frac{1}{5}$ are the roots of the following quadratic trinomial
$$100x^2 - 25x + 1.$$
Therefore, the set of the solutions of the following inequality
$$100x^2 - 25x + 1 < 0,$$
is the interval $\left(\frac{1}{20}, \frac{1}{5}\right)$. We need to find the total number of all positive integers, such that
$$\frac{1}{20} < \frac{1}{n} < \frac{1}{5}.$$
Hence, we deduce that
$$5 < n < 20.$$
Thus, it follows that
$$n \in \{6, 7, ..., 19\}.$$
Therefore, the total number of all such elements of the given set is equal to 14. $\square$

Problem 4.206. *A student took the bus to Harvard from Newton. He got into the bus at the first station and got off at the last station. The student noticed, that 10% of all passengers got into the bus at the first station and 60% of all passengers got off at the last station. The student also noticed, that in between the first and the last stations 8 people got off the bus. How many people got into the bus at the first station?*

(A) 1   (B) 3   (C) 10   (D) 5   (E) 2

Solution. Answer. (E)
According to the assumption of the problem, we have that in between the first and the last stations 40% of all passengers got off the bus.
Hence, 40% of all passengers is equal to 8. Thus, it follows that 10% is equal to 2.
Therefore, 2 people got into the bus at the first station. $\square$

Problem 4.207. *For which value of m do lines $y = x - 1$, $y = 3x - 5$ and $y = mx - 41$ intersect at one point?*

(A) 5   (B) 10   (C) 19   (D) 21   (E) 22

Solution. Answer. (D)
At first, let us find the intersection point $M(x_0, y_0)$ of the lines $y = x - 1$ and $y = 3x - 5$.
Note that $y_0 = x_0 - 1$ and $y_0 = 3x_0 - 5$.
Therefore $x_0 - 1 = 3x_0 - 5$.
Hence, we obtain that $x_0 = 2$ and $y_0 = x_0 - 1 = 1$.
According to the assumption of the problem, we have that the line $y = mx41$ passes through the point $M(2, 1)$.
Thus, it follows that $1 = 2m41$. Therefore $m = 21$. □

Problem 4.208. *One of the shops sells 450 grams of a candy for 5\$, the second shop sells 500 grams of the same candy for 6\$. By how many percent the candy sold in the second shop is more expensive than the candy sold in the first shop?*

(A) 12   (B) 10   (C) 9   (D) 8   (E) 5

Solution. Answer. (D)
According to the assumption of the problem, we have that the first shop sells 50 grams of a candy for $\frac{5}{9}$\$.
On the other hand, the second shop sells 50 grams of the same candy for $\frac{3}{5}$\$.
Note that $\frac{3}{5}$ is greater than $\frac{5}{9}$ by the following percent:

$$\frac{\frac{3}{5} - \frac{5}{9}}{\frac{5}{9}} \cdot 100 = 8.$$

□

Problem 4.209. *Let a and b be real numbers, such that $a + b = 1$ and $(a^2 + b)(b^2 + a) = 2019$. What is the value of the expression $(a^2 + 1)(b^2 + 1)$?*

(A) 100   (B) 200   (C) 2018   (D) 2019   (E) 2020

Solution. Answer. (E)
Given that
$$(a^2 + b)(b^2 + a) = 2019.$$
Thus, it follows that
$$a^2 b^2 + ab + a^3 + b^3 = 2019.$$
Given also that
$$a + b = 1.$$
Taking into consideration that
$$a^3 + b^3 = (a + b)(a^2 - ab + b^2),$$
we obtain that
$$a^2 b^2 + ab + a^2 - ab + b^2 = 2019.$$
Hence, we deduce that
$$(a^2 + 1)(b^2 + 1) = 2020.$$

□

173

Problem 4.210. *Let D be a given point on side AC triangle ABC, such that $AD = 1$, $CD = 7$, $BD = 4$ and $\angle ADB = 90°$. What is the length of the median drawn from vertex B of triangle ABC?*

(A) $3\sqrt{2}$ (B) $2\sqrt{5}$ (C) 5 (D) $3\sqrt{3}$ (E) 6

Solution. Answer. (C)
Let M be the midpoint of side AC (see the figure).

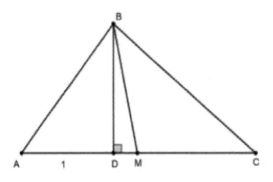

Note that
$$AM = \frac{AC}{2} = 4.$$
On the other hand, we have that
$$DM = AM - AD = 3.$$
According to Pythagorean theorem, from right triangle BDM, we obtain that
$$BM^2 = BD^2 + DM^2.$$
Thus, it follows that
$$BM = 5.$$
$\square$

Problem 4.211. *Let $\overline{ab}$, $\overline{cd}$, $\overline{ac}$, $\overline{bd}$ be two-digit numbers, such that $\overline{ab} + \overline{cd} = \overline{ac} + \overline{bd}$. What is the total number of all possible quadruples (a, b, c, d)?*

(A) 810 (B) 729 (C) 700 (D) 500 (E) 100

Solution. Answer. (A)
Given that
$$\overline{ab} + \overline{cd} = \overline{ac} + \overline{bd}.$$
Thus, it follows that
$$10a + b + 10c + d = 10a + c + 10b + d.$$
Hence, we obtain that $c = b$.
Thus, we need to find the total number of all possible quadruples (a, b, c, d), such that
$$a \in \{1, 2, ..., 9\}, b \in \{1, 2, ..., 9\}, d \in \{0, 1, 2, ..., 9\}.$$
Therefore, the total number of all such quadruples is eqaul to 810, as $9 \cdot 9 \cdot 10 = 810$. $\square$

Problem 4.212. *Given three pairwise different positive odd integers, such that the sum of any two of them is greater at least by 5 than the third number. What is the smallest possible value of the sum of these three numbers?*

(A) 17   (B) 23   (C) 25   (D) 27   (E) 29

Solution. Answer. (D)
Let given three numbers be a, b, c. Without loss of generality one can assume that
$$a < b < c.$$
According to the assumption of the problem, we have that
$$a + b \geq c + 5.$$
Thus, it follows that
$$a \geq (c - b) + 5 \geq 2 + 5 = 7.$$
Hence, we obtain that $a \geq 7$.
Therefore $b \geq 9$ and $c \geq 11$.
We deduce that
$$a + b + c \geq 27.$$
Note that numbers 7, 9, 11 satisfy this condition and the assumptions of the problem.
Therefore, the smallest possible value of the sum of these three numbers is equal to 27. □

Problem 4.213. *Let two circles of radii 6 be tangent to each other, such that each of these circles is tangent to bases AD and BC and to one of the legs of trapezoid ABCD. Given that $AB = 13$ and $CD = 20$. What is the area of trapezoid ABCD?*

(A) 198   (B) 396   (C) 300   (D) 342   (E) 210

Solution. Answer. (D)
Let us consider the figure below.

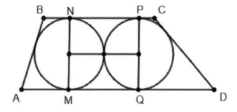

We have that
$$Area(ABCD) = Area(ABNM) + Area(MNPQ) + Area(QPCD) =$$
$$= Area(ABNM) + Area(QPCD) + 12 \cdot 12.$$
Note that, if we connect trapezoids $ABNM$ and $PCDQ$, such that line segments MN and PG coincide, then the area of the obtained trapezoid is equal to:
$$(AB + CD) \cdot 6 = 198.$$
Hence, we obtain that
$$Area(ABNM) + Area(QPCD) = 198.$$
Thus, it follows that
$$Area(ABCD) = 198 + 144 = 342.$$
□

Problem 4.214. *What is the value of the following expression?*

$$\frac{3}{1\cdot 2} - \frac{5}{2\cdot 3} + \frac{7}{3\cdot 4} - \frac{9}{4\cdot 5} + \frac{11}{5\cdot 6} - \frac{13}{6\cdot 7} + \frac{15}{7\cdot 8} - \frac{17}{8\cdot 9} + \frac{19}{9\cdot 10}.$$

(A) 1.1   (B) 0.9   (C) 0.75   (D) 0.6   (E) 0.5

Solution. Answer. (A)
Note that for any positive integer k we have that

$$\frac{2k+1}{k(k+1)} = \frac{(k+1)+k}{k(k+1)} = \frac{k+1}{k(k+1)} + \frac{k}{k(k+1)} = \frac{1}{k} + \frac{1}{k+1}.$$

Thus, it follows that for any positive integer k we have that

$$\frac{2k+1}{k(k+1)} = \frac{1}{k} + \frac{1}{k+1}.$$

Taking this into consideration, let us rewrite the given expression in the following way

$$\frac{3}{1\cdot 2} - \frac{5}{2\cdot 3} + \frac{7}{3\cdot 4} - \frac{9}{4\cdot 5} + \frac{11}{5\cdot 6} - \frac{13}{6\cdot 7} + \frac{15}{7\cdot 8} - \frac{17}{8\cdot 9} + \frac{19}{9\cdot 10} =$$

$$= \frac{1}{1} + \frac{1}{2} - \left(\frac{1}{2} + \frac{1}{3}\right) + \frac{1}{3} + \frac{1}{4} - \left(\frac{1}{4} + \frac{1}{5}\right) + \frac{1}{5} + \frac{1}{6} - \left(\frac{1}{6} + \frac{1}{7}\right) +$$

$$+ \frac{1}{7} + \frac{1}{8} - \left(\frac{1}{8} + \frac{1}{9}\right) + \frac{1}{9} + \frac{1}{10} = 1 + \frac{1}{10} = 1.1.$$

□

Problem 4.215. *The numbers -5, -4, -3, -2, -1, 0, 1, 2, 3, 4, 5 are written on the board. James can pick any two of these numbers, for example numbers a and b, afterward James writes the value of the expression $ab - a - b + 2$ on the board. Then he erases numbers a and b from the board. For the next step, James repeats the same with the remaining 10 numbers. After the 10th step, only one number is written on the board. What is the value of the last number left on the board?*

(A) -120   (B) -5   (C) 0   (D) 1   (E) 4

Solution. Answer. (D)
Note that if the last number left on the board is 1, then the answer is 1. Otherwise, according to the assumption of the problem, we have that whenever James erases numbers 1 and x, instead of them he writes on the board the expression $1 \cdot x - 1 - x + 2 = 1$. Therefore, after this step numbers 1 and x get replaced by number 1. In a similar way, if afterward James chooses number 1 with any other number, then after that step also he replaces these two chosen numbers by 1.
From the above mentioned it follows that the last number left on the board is equal to 1.   □

Problem 4.216. *Let m be a positive integer and n be a non-negative integer. What is the total number of all (m, n) pairs, such that*

$$\sum_{i=1}^{m!} i = 2^{2n} + 2^n + 1?$$

(A) 1   (B) 2   (C) 3   (D) 6   (E) 7

Solution. Answer. (B)
Note that $2^{2n} + 2^n + 1$ is an odd number and
$$\sum_{i=1}^{m!} i = \frac{m!}{2}(m! + 1).$$
Thus, it follows that $\frac{m!}{2}(m! + 1)$ is an odd number.

Hence, we obtain that $m \leq 3$. A straightforward verification shows that pairs $(2, 0)$ and $(3, 2)$ satisfy the assumptions of the problem.

Therefore, the total number of all such (m, n) pairs is equal to 2. □

Problem 4.217. *Let a and b be positive integers and $a > b$. Consider points $A(a, b)$, $B(a - b, a)$, $O(0, 0)$ on the coordinate plane. Given that the area of triangle ABO is 96. What is the value of $a + b$?*

(A) 28   (B) 27   (C) 26   (D) 25   (E) 24

Solution. Answer. (E)
Let us consider the figure below.

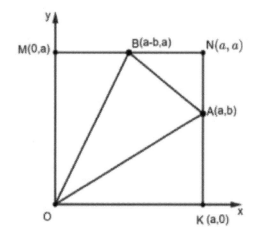

Note that
$$(OAB) = (OMNK) - (OMB) - (BNA) - (OAK) =$$
$$= a^2 - \frac{a(a-b)}{2} - \frac{b(a-b)}{2} - \frac{ab}{2} = \frac{a^2 - ab + b^2}{2}.$$
According to the assumption of the problem, we have that
$$a^2 - ab + b^2 = 192.$$
Thus, it follows that
$$a^2 + b^2 + (a-b)^2 = 384 = 3 \cdot 2^7.$$
Note that the square of any odd number leaves a remainder of 1 after division by 4. Hence, we obtain that
$$a = 16, b = 8.$$
Thus, it follows that
$$a + b = 24.$$
□

Problem 4.218. *Let AD is the bisector of angle BAC of triangle ABC. Given that $BD = 3$, $CD = 5$ and $AC - AB = 4$. What is the value of the measure (in degrees) of angle ABC?*

(A) 120 (B) 100 (C) 90 (D) 75 (E) 60

Solution. Answer. (C)
Let us choose point E on side AC, such that $AE = AB$ (see the figure).

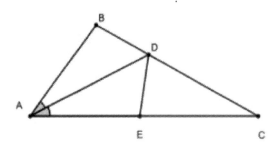

According to the assumption of the problem, we have that
$$EC = AC - AE = AC - AB = 4.$$

Note that $\triangle ABD$ is congruent to $\triangle AED$. Thus, it follows that $BD = DE$ and $\angle ABD = \angle AED$. Hence, we deduce that
$$DE = 3, EC = 4, CD = 5.$$

Therefore, according to the converse of Pythagorean theorem from triangle DEC we obtain that $\angle DEC = 90°$. Thus, it follows that
$$\angle ABC = \angle ABD = \angle AED = 180° - \angle DEC = 90°.$$

□

Problem 4.219. *What is the total number of all triples (p, q, r) of prime numbers, such that $p < q < r$ and each of the numbers $q - p$, $r - q$, $r - p$ is not a composite number?*

(A) 0 (B) 1 (C) 2 (D) 6 (E) 12

Solution. Answer. (C)
Let us prove that $p = 2$.
We proceed the proof by contradicition argument. If $2 < p$, then $r - p \geq 4$ and $r - p$ is an even number. Thus, it follows that $r - p$ is a composite number. This leads to a contradicition, as according to the assumption of the problem $r - p$ is a prime number. Therefore $p = 2$.
If $q = 3$, then note that $r = 5$. Otheriwse, if $r > 5$, then $r - q$ is not a composite number.
When $q > 3$, then some two of the numbers p, q, r leave the same remainder after division by 3. According to the assumption of the problem, the positive difference of those two numbers is equal to 3. Therefore, $q = 5$ and $r = 7$.
Note that triples of prime numbers $(2, 3, 5)$ and $(2, 5, 7)$ satisfy assumptions of the problem.   □

Problem 4.220. *Given a rectangle $ABCD$ and circles σ_1, σ_2, σ_3, σ_4, each with radius $r = \sqrt{2-\sqrt{3}}$ (see the figure). Let each of the circles σ_1, σ_3 touch each of the circles σ_2, σ_4. Given that circles σ_2, σ_4 touch each other and sides AB, CD, respectively. Given also that circles σ_1, σ_3 touch respectively sides AB, AD and BC, CD. What is the area of rectangle $ABCD$?*

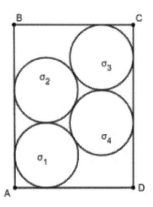

(A) $2\sqrt{3}(2-\sqrt{3})$ (B) 4 (C) $2\sqrt{3}$ (D) 5 (E) $3+2\sqrt{3}$

Solution. Answer. (D)
Let O_1, O_2, O_3, O_4 be the centers of circles σ_1, σ_2, σ_3, σ_4, respectively (see the figure).

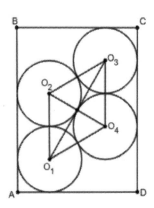

According to the assumption of the problem, we have that $O_1O_2O_3O_4$ is a rhombus with side length $2\sqrt{2-\sqrt{3}}$. Moreover $\angle O_2O_1O_4 = 60°$ and $O_1O_2 \parallel AB$. Note that

$$AD = r + \frac{O_1O_3}{2} + r = (2+\sqrt{3})r,$$

and

$$AB = r + O_1O_3 \cos 30° + r = 5r.$$

Thus, it follows that

$$\text{Area}(ABCD) = 5r(2+\sqrt{3})r = 5(2+\sqrt{3})(2-\sqrt{3}) = 5.$$

□

Problem 4.221. *What is the total number of all seven-digit numbers consisting only of the digits 1, 2, 3, 4, 5, 6, 7 (in any order), such that the digits are not repetitive and each is divisible by 11?*

(A) 120   (B) 240   (C) 320   (D) 400   (E) 576

Solution. Answer. (D)
Let $\overline{abcdefg}$ be a seven-digit number and it satisfies the assumptions of the problem. We have that

$$11 \mid (a - b + c - d + e - f + g).$$

On the other hand, we have that

$$a - b + c - d + e - f + g = a + b + c + d + e + f + g - 2(b + d + f) =$$
$$= 1 + 2 + 3 + 4 + 5 + 6 + 7 - 2(b + d + f) = 2(14 - (b + d + f)).$$

Thus, it follows that
$$b + d + f = 14.$$

Note that the following cases are possible:

$$\{b, d, f\} = \{1, 6, 7\},$$
$$\{b, d, f\} = \{2, 5, 7\},$$
$$\{b, d, f\} = \{3, 4, 7\},$$
$$\{b, d, f\} = \{3, 5, 6\}.$$

Therefore, the total number of seven-digit numbers satisfying the assumption of the problem is equal to 576, as $4 \cdot 3! \cdot 4! = 576$. □

Problem 4.222. *Let $\overline{abc}$ be the smallest three-digit number, such that when it is divided by the product of its digits the remainder is equal to the sum of its digits. What is the quotient of that division?*

(A) 11   (B) 12   (C) 13   (D) 14   (E) 15

Solution. Answer. (D)
According to the assumption of the problem, we have that

$$\overline{abc} = a \cdot b \cdot c \cdot q + a + b + c,$$

and
$$abc > a + b + c.$$

Let us choose $a = 1$, then from the condition $abc > a + b + c$ follows that $b \neq 1$.
Note also that from the condition $9(11 + b) = b \cdot c \cdot q$ follows that $b \neq 2$. Hence, we obtain that $b \geq 3$.
If $b = 3$, then we have that $1 \cdot 3 \cdot c > 1 + 3 + c$. Therefore $c > 2$. Thus, it follows that

$$\overline{abc} \geq 133.$$

If $a \geq 2$, then again we obtain that
$$\overline{abc} \geq 133.$$

On the other hand, 133 satisifes the assumptions of the problem, as

$$133 = 1 \cdot 3 \cdot 3 \cdot 14 + 1 + 3 + 3.$$

Therefore, the quotient of the given division is equal to 14. □

Problem 4.223. *Points M, N, P, Q are given correspondingly on sides AB, BC, CD, AD of the rectangle $ABCD$. Let R be the intersection point of line segments MP and QN. Given that $\angle MRN = 90°$, $MR = 21$, $NR = 27$, $QR = 33$ and $\dfrac{AB}{BC} = \dfrac{3}{4}$. What is the length of PR?*

(A) 39 (B) $42\dfrac{3}{7}$ (C) 45 (D) 59 (E) 61

Solution. Answer. (D)
Let $ME \perp CD$ and $NF \perp AD$ (see the figure).

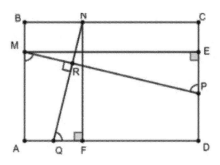

Note that
$$\angle MPE = \angle AMP = 360° - (\angle MAQ + \angle MRQ + \angle AQR) =$$
$$= 180° - \angle AQR = \angle NQF.$$

Thus, we obtain that
$$\angle MPE = \angle NQF.$$

Hence, we obtain that $\triangle QNF \sim \triangle PME$. Therefore
$$\frac{QN}{PM} = \frac{NF}{ME}.$$

Note that $MBCE$ and $ABNF$ are rectangles, thus
$$ME = BC, NF = AB.$$

From the above mentioned two equations, we deduce that
$$\frac{QN}{PM} = \frac{NF}{ME} = \frac{AB}{BC} = \frac{3}{4}.$$

Hence, we obtain that
$$PM = \frac{4}{3}QN = \frac{4}{3}(QR + NR) = 80.$$

Thus, it follows that
$$MR + PR = 80.$$

Hence, we deduce that
$$PR = 80 - 21 = 59.$$

□

Problem 4.224. *There were 8 chairs at a round table and on each chair was sitting one person. After the break, each person randomly chooses to sit either on the chair that he/she was sitting before the break or on one of its neighboring two chairs (again on each chair is sitting one person). What is the probability of the event that after the break exactly two persons are sitting on their previous chairs?*

(A) $\dfrac{1}{3}$ (B) $\dfrac{8}{39}$ (C) $\dfrac{1}{2}$ (D) $\dfrac{16}{53}$ (E) $\dfrac{16}{49}$

Solution. Answer. (E)
Let a_n be the number of possible ways that after the break exactly n people were sitting on their previous chairs.
Let us enumerate (towards one direction) all the chairs with the numbers 1, ..., 8.
$i \to j$ means that before the break the person who was sitting on the ith chair, after the break is sitting on the jth chair.
Let us prove that $a_0 = 4$. Note that $1 \to 2$ or $1 \to 8$.
If $1 \to 2$ and $2 \to 3$, then $3 \to 4$, $4 \to 5$,..., $7 \to 8$, $8 \to 1$.
If $1 \to 2$ and $2 \to 1$, then $3 \to 4$, $4 \to 3$, otherwise $8 \to 1$ (note that this is not correct).
In a similar way, we obtain that $5 \to 6$, $6 \to 5$, $7 \to 8$, $8 \to 7$.
We obtain that when $1 \to 2$, then the number of the ways is equal to 2. In a similar way, we obtain that when $1 \to 8$, then again the number of ways is equal to 2. Therefore $a_0 = 4$.
Let us prove that
$$a_1 = a_3 = a_5 = a_7 = 0.$$
This holds true due to the fact that: if $i \to i$, then $i+1 \to i+1$ or $i+1 \to i+2$ and $i+2 \to i+1$, where $i \in \{1, ..., 8\}$ and the chairs with numbers 9, 10 are correspondingly the chairs with numbers 1,2.
Using the above mentioned property, we obtain that
$$a_2 = 8 + 8 = 16,$$
$$a_4 = 8 + 8 + 4 = 20,$$
$$a_6 = 8, a_8 = 1.$$

Thus, it follows that the probability of the event that after the break exactly two persons are sitting on their previous chairs is:
$$\frac{a_2}{\sum_{i=0}^{8} a_i} = \frac{16}{49}.$$
$\square$

Problem 4.225. *What is the value of the sum of all positive integers n, such that for each such n the value of the expression $\dfrac{n(n+7)}{20}$ is equal to some prime number raised to a positive integer power?*

(A) 18 (B) 33 (C) 38 (D) 61 (E) 66

Solution. Answer. (A)
According to the assumption of the problem let n be a positive integer, such that
$$\frac{n(n+7)}{20} = p^m,$$
where p is a prime number and m is a positive integer.
Let us consider the following two cases.
Case 1. If $p \neq 7$.
Note that
$$20 \mid n(n+7),$$

182

and
$$p \mid n(n+7).$$

On the other hand, the statement $n + 7 - n = 7$ implies that only one of the numbers n and $n + 7$ is divisible by 2.

In a similar way, we obtain that only one of these numbers is divisible by 5 and p.

Therefore, either $n = 4$ or $n = 5$ or $n = 20$ or $n + 7 = 20$.

Note that numbers 5, 13, 20 satisfy the assumptions of the problem.

Case 2. If $p = 7$.

We have that
$$\frac{n(n+7)}{20} = 7^m.$$

Thus, it follows that $n = 7k$, where k is a positive integer.

Hence, we deduce that
$$\frac{k(k+1)}{20} = 7^{m-2}.$$

Therefore, either $k = 4$ or $k + 1 = 4$ or $k = 5$ or $k + 1 = 5$ or $k = 20$ or $k + 1 = 20$.

Note that from these options only $k = 4(n = 28)$ satisfies the assumption of the problem.

Therefore, the sum of all such positive integers n is equal to 66, as $5 + 13 + 20 + 28 = 66$. □

4.10 Solutions of AMC 10 type practice test 10

Problem 4.226. *What is the value of the following expression?*

$$\sqrt[6]{2^5 \cdot \sqrt[3]{2^6 \cdot \sqrt[8]{2^7 \cdot \sqrt[9]{2^8 \cdot \sqrt[10]{2^{10}}}}}}.$$

(A) $\sqrt[6]{2}$ (B) 2   (C) 4   (D) 8   (E) 1024

Solution. Answer. (B)
Note that

$$\sqrt[6]{2^5 \cdot \sqrt[7]{2^6 \cdot \sqrt[8]{2^7 \cdot \sqrt[9]{2^8 \cdot \sqrt[10]{2^{10}}}}}} = \sqrt[6]{2^5 \cdot \sqrt[7]{2^6 \cdot \sqrt[8]{2^7 \cdot \sqrt[3]{2^8 \cdot 2}}}} =$$

$$= \sqrt[6]{2^5 \cdot \sqrt[7]{2^6 \cdot \sqrt[8]{2^7 \cdot 2}}} = \sqrt[6]{2^5 \cdot \sqrt[7]{2^6 \cdot 2}} = \sqrt[6]{2^6} = 2.$$

□

Problem 4.227. *The arithmetic mean of $a, 3, 4$ is 5 more than the arithmetic mean of $b, -1, 4$. What is the value of non-negative difference of a and b?*

(A) 18   (B) 5   (C) 11   (D) 0   (E) 2

Solution. Answer. (B)
According to the assumption of the problem, we have that

$$\frac{a+3+4}{3} - \frac{b+(-1)+4}{3} = 5.$$

Thus, it follows that
$$a + 7 - (b+3) = 15.$$

Therefore, we obtain that
$$a - b = 15 - 4 = 11.$$

□

Problem 4.228. *What is the value of x, such that the following equation holds true?*

$$x - \frac{1}{6} = \frac{1}{1 \cdot 2} + \frac{1}{2 \cdot 3} + \frac{1}{3 \cdot 4} + \frac{1}{4 \cdot 5} + \frac{1}{5 \cdot 6}.$$

(A) 0   (B) $\frac{1}{6}$ (C) $\frac{31}{30}$ (D) $\frac{1}{2}$ (E) 1

Solution. Answer. (E)
Note that

$$\frac{1}{1 \cdot 2} + \frac{1}{2 \cdot 3} + \frac{1}{3 \cdot 4} + \frac{1}{4 \cdot 5} + \frac{1}{5 \cdot 6} = \frac{2-1}{1 \cdot 2} + \frac{3-2}{2 \cdot 3} + \frac{4-3}{3 \cdot 4} + \frac{5-4}{4 \cdot 5} + \frac{6-5}{5 \cdot 6} =$$

$$= \frac{2}{1 \cdot 2} - \frac{1}{1 \cdot 2} + \frac{3}{2 \cdot 3} - \frac{2}{2 \cdot 3} + \frac{4}{3 \cdot 4} - \frac{3}{3 \cdot 4} + \frac{5}{4 \cdot 5} - \frac{4}{4 \cdot 5} + \frac{6}{5 \cdot 6} - \frac{5}{5 \cdot 6} =$$

$$= 1 - \frac{1}{2} + \frac{1}{2} - \frac{1}{3} + \frac{1}{3} - \frac{1}{4} + \frac{1}{4} - \frac{1}{5} + \frac{1}{5} - \frac{1}{6} = 1 - \frac{1}{6}.$$

□

Problem 4.229. *What is the value of the smallest positive integer n, such that the value of the expression $\dfrac{n-2}{5} - \dfrac{n+3}{6}$ is also a positive integer?*

(A) 1 (B) 10 (C) 27 (D) 57 (E) 87

Solution. Answer. (D)
Note that
$$\frac{n-2}{5} - \frac{n+3}{6} = \frac{n-27}{30}.$$
Thus, it follows that $n > 27$ and $30 \mid (n-27)$.
Hence, we obtain that the smallest possible value of n is equal to 57. □

Problem 4.230. *John got 150000\$ 30-year fixed 4.8 % interest rate mortgage loan from the bank. Each year in total he should pay back the same amount of money (principal+interest) and each year 4.8% interest is calculated on the principal left after the payment of the previous year. In total how much money should John pay to the bank in the end of the second year?*

(A) 11625 (B) 11960 (C) 12000 (D) 15000 (E) 54800

Solution. Answer. (B)
Note that each year John should return to the bank 5000\$ of principal of the mortgage loan, as
$$\frac{150000}{30} = 5000.$$
In one year, from principal of the loan will be left 145,000 to be paid, as $150000 - 5000 = 145000$. Therefore, in the second year John should pay an interest of 6960\$, as
$$\frac{145000 \cdot 4.8}{100} = 6960.$$
Therefore, in the end of the second year in total John needs to pay to the back 11960\$, as $5000 + 6960 = 11960$. □

Problem 4.231. *What is the value of the following expression?*
$$\left(\frac{1}{7} + \frac{1}{21} - \frac{1}{14} - \frac{1}{28} - \frac{1}{12}\right)^2 + \left(\frac{1}{5} + \frac{1}{10} + \frac{1}{20} - \frac{1}{4} - \frac{1}{15} - \frac{1}{30}\right)^2 + \left(\frac{1}{8} + \frac{1}{24} - \frac{1}{9} - \frac{1}{18}\right)^2 + \left(\frac{1}{2} + \frac{1}{3} + \frac{1}{6} - 1\right)^2.$$

(A) $-\dfrac{1}{2}$ (B) $-\dfrac{1}{3}$ (C) 0 (D) $\dfrac{1}{3}$ (E) 1

Solution. Answer. (C)
Note that
$$\frac{1}{7} + \frac{1}{21} - \frac{1}{14} - \frac{1}{12} - \frac{1}{28} = 0,$$
$$\frac{1}{5} + \frac{1}{10} + \frac{1}{20} - \frac{1}{4} - \frac{1}{15} - \frac{1}{30} = 0,$$
$$\frac{1}{8} + \frac{1}{24} - \frac{1}{9} - \frac{1}{18} = 0,$$
$$\frac{1}{2} + \frac{1}{3} + \frac{1}{6} = 1.$$
Thus, it follows that
$$\left(\frac{1}{7} + \frac{1}{21} - \frac{1}{14} - \frac{1}{28} - \frac{1}{12}\right)^2 + \left(\frac{1}{5} + \frac{1}{10} + \frac{1}{20} - \frac{1}{4} - \frac{1}{15} - \frac{1}{30}\right)^2 + \left(\frac{1}{8} + \frac{1}{24} - \frac{1}{9} - \frac{1}{18}\right)^2 + \left(\frac{1}{2} + \frac{1}{3} + \frac{1}{6} - 1\right)^2 = 0.$$
□

Problem 4.232. *What is the value of the sum of all distinct solutions of the equation $((x-4)^2-3)^2 = 9$?*

(A) 16   (B) 15   (C) 14   (D) 12   (E) 0

Solution. Answer. (D)
Note that the given equation is equivalent to the following equation
$$((x-4)^2 - 3 - 3)((x-4)^2 - 3 + 3) = 0.$$
Thus, it follows that
$$(x^2 - 8x + 10)(x-4)^2 = 0.$$
Note that all distinct solutions of the last equation are 4, $4 - \sqrt{6}$, $4 + \sqrt{6}$. Therefore, the sum of all distinct roots of the given equation is equal to 12, as $4 + 4 - \sqrt{6} + 4 + \sqrt{6} = 12$. □

Problem 4.233. *Two numbers are randomly chosen from the set of all two-digit numbers. What is the probability that the positive difference of two chosen numbers is also a two-digit number?*

(A) $\dfrac{36}{89}$ (B) $\dfrac{2}{5}$ (C) $\dfrac{72}{89}$ (D) $\dfrac{4}{5}$ (E) $\dfrac{9}{10}$

Solution. Answer. (C)
Note that the number of all possible cases is equal to
$$\binom{90}{2} = 45 \cdot 89.$$
Let us assume that the positive difference of two chosen numbers is a two-digit number.
Let the smaller number from two chosen numbers be x and let the number of options to choose the bigger number be $n(x)$.
We have that $n(x) = 0$, if $x \in \{90, 91, ..., 99\}$. We also have that $n(x) = 90 - x$, if $x \in \{10, 11, ..., 89\}$. Therefore, the number of all favorable cases is equal to
$$1 + 2 + ... + 80 = 40 \cdot 81.$$
Thus, it follows that the probability that the positive difference of two chosen numbers is also a two-digit number is
$$\frac{40 \cdot 81}{45 \cdot 89} = \frac{72}{89}.$$
□

Problem 4.234. *A rectangular prism is called "beautiful" if its three measurements are positive integers. The rectangular prism M is divided into eight "beautiful" rectangular prisms with three planes parallel to its faces. Given that out of these eight "beautiful" rectangular prisms the volumes of four of them are equal to 1, 2, 3, and 5 What is the surface area of the rectangular prism M?*

(A) 108   (B) 84   (C) 72   (D) 60   (E) 30

Solution. Answer. (A)
According to the assumption of the problem, we have that the sizes of four of those eight "beautiful" rectangles are $1 \times 1 \times 1$, $1 \times 1 \times 2$, $1 \times 1 \times 3$ and $1 \times 1 \times 5$.
Therefore, the size of rectangle M is $3 \times 4 \times 6$. Hence, we obtain that the surface area of rectangle M is equal to:
$$2(3 \cdot 4 + 4 \cdot 6 + 3 \cdot 6) = 108.$$
□

Problem 4.235. Let number a be the smallest value of variable x, such that the values of functions $y = \frac{2}{3}x - \frac{1}{3}$ and $y = \frac{3}{4}x - \frac{1}{4}$ are positive integers. What is the value of the sum of all digits of a?

(A) 2   (B) 3   (C) 4   (D) 1   (E) 5

Solution. Answer. (A)
Let m and n be positive integers, such that
$$\frac{2}{3}a - \frac{1}{3} = m,$$
and
$$\frac{3}{4}a - \frac{1}{4} = n.$$
Thus, it follows that
$$a = \frac{3}{2}m + \frac{1}{2},$$
and
$$a = \frac{4}{3}n + \frac{1}{3}.$$
Hence, we deduce that
$$\frac{3}{2}m + \frac{1}{2} = \frac{4}{3}n + \frac{1}{3}.$$
We obtain that
$$9m + 3 = 8n + 2.$$
Thus, it follows that
$$9m + 1 = 8n.$$
Note that a is the smallest if positive integer m is the smallest.
Note also that the smallest positive integer value of m is equal to 7. Therefore $a = 11$ and the sum of all digits of a is equal to 2. □

Problem 4.236. Steven planned to solve some number of problems in 3 days. On the first day, he solved $\frac{1}{3}$ of all problems. On the second and the third days, he solved respectively $\frac{3}{4}$ and $\frac{5}{6}$ of problems. It turned out that the number of problems he solved was 6 times more than the number of problems he did not solved. What percent of all problems did Steven solve on the second day?

(A) 25   (B) $\frac{200}{7}$ (C) $\frac{300}{7}$ (D) 50   (E) 51

Solution. Answer. (B)
Let the entire assignment be 1 unit and the assignment for the second day be x unit. The part of the assignment planned for the third day is $\frac{2}{3} - x$.
According to the assumption of the problem, we have that
$$\frac{1}{3} + \frac{3}{4}x + \frac{5}{6}\left(\frac{2}{3} - x\right) = 6\left(\frac{1}{4}x + \frac{1}{6}\left(\frac{2}{3} - x\right)\right).$$
Thus, it follows that
$$\frac{1}{3} = \frac{3}{4}x + \frac{1}{6}\left(\frac{2}{3} - x\right).$$
Hence, we obtain that
$$x = \frac{8}{21}.$$

We deduce that
$$\frac{3}{4}x = \frac{2}{7}.$$
Therefore, on the second day Steven has solved $\frac{200}{7}$ percent of all problems. □

Problem 4.237. *Given a triangle ABC. Let CD be the angle bisector of ∠ACB. Given that $AC = CD$, $\angle ACB = 108°$ and $\angle A = n \cdot \angle B$. What is the value of n?*

(A) 3 (B) 5 (C) 6 (D) 7 (E) 10

Solution. Answer. (D)
We have that
$$\angle ACD = \frac{1}{2}\angle C = 54°,$$
and
$$\angle CAD = \angle CDA = \frac{180° - 54°}{2} = 63°.$$
Thus, it follows that
$$\angle A = 63°,$$
and
$$\angle B = 180° - (108° + 63°) = 9^{circ}.$$
Taking this into consideration and using that $\angle A = n \cdot \angle B$, we obtain that $n = 7$. □

Problem 4.238. *A positive divisor of 10! is randomly chosen. What is the probability that the chosen divisor is not divisible by 3?*

(A) $\frac{2}{3}$ (B) $\frac{1}{3}$ (C) $\frac{1}{4}$ (D) $\frac{1}{5}$ (E) $\frac{1}{6}$

Solution. Answer. (D)
Let the total number of all positive divisors of 10! be n.
Note that $10! = 3^4 \cdot m$, where m is a positive integer not divisible by 3.
Therefore, all the positive divisors of 10! that are not divisible by 3 is equal to $\frac{n}{5}$.
Thus, it follows that the probability that the chosen divisor of 10! is not divisible by 3 is equal to $\frac{1}{5}$, as
$$\frac{\frac{n}{5}}{n} = \frac{1}{5}.$$
□

Problem 4.239. *Let $P(x) = (x-2)(x-4)(x-6)(x-8)$. What is the total number of all possible pairs of integers (m, n), such that $P(m) + P(n) < 0$.*

(A) 36 (B) 35 (C) 34 (D) 33 (E) 32

Solution. Answer. (E)
Note that
$$P(x) = P(10 - x).$$
Besides, we have that
$$P(2) = P(4) = P(6) = P(8) = 0,$$
$$P(3) = P(7) = -15, P(5) = 9,$$

and if k is an integer not less than 9, then
$$P(k) \geq 7 \cdot 5 \cdot 3 \cdot 1 = 105.$$

Therefore, if $k \leq 1$, then $10 - k \geq 9$.
Thus, it follows that
$$P(k) = P(10 - k) \geq 105.$$

Note also that $P(m) + P(n) < 0$ holds true, either if $m \in \{3, 7\}$ and $n \in \{2, 3, 4, 5, 6, 7, 8, 9, 10\}$, or $n \in \{3, 7\}$ and $min\{2, 3, 4, 5, 6, 7, 8, 9, 10\}$.
Therefore, the number of (m, n) pairs is $2 \cdot 2 \cdot 9 - 4 = 32$. □

Problem 4.240. *What is the value of the following expression?*

$$\frac{1^2 + 5 \cdot 1 + 4}{1^2 + 5 \cdot 1 + 6} \cdot \frac{2^2 + 5 \cdot 2 + 4}{2^2 + 5 \cdot 2 + 6} \cdot \ldots \cdot \frac{98^2 + 5 \cdot 98 + 4}{98^2 + 5 \cdot 98 + 6}.$$

(A) 0.35   (B) 0.4   (C) 0.5   (D) 0.51   (E) 0.61

Solution. Answer. (D)
Let us consider and factorize the trinomials $x^2 + 5x + 4$ and $x^2 + 5x + 6$.
We have that
$$x^2 + 5x + 4 = x[2] + x + 4x + 4 = (x + 1)(x + 4),$$
and
$$x^2 + 5x + 6 = x^2 + 2x + 3x + 6 = (x + 2)(x + 3).$$

Taking this into consideration, we obtain that

$$\frac{1^2 + 5 \cdot 1 + 4}{1^2 + 5 \cdot 1 + 6} \cdot \frac{2^2 + 5 \cdot 2 + 4}{2^2 + 5 \cdot 2 + 6} \cdot \ldots \cdot \frac{98^2 + 5 \cdot 98 + 4}{98^2 + 5 \cdot 98 + 6} =$$

$$= \frac{2 \cdot 5}{3 \cdot 4} \cdot \frac{3 \cdot 6}{4 \cdot 5} \cdot \frac{4 \cdot 7}{5 \cdot 6} \cdot \ldots \cdot \frac{99 \cdot 102}{100 \cdot 101} =$$

$$\frac{2 \cdot 3 \cdot 4 \cdot \ldots \cdot 99}{3 \cdot 4 \cdot 5 \cdot \ldots \cdot 100} \cdot \frac{5 \cdot 6 \cdot 7 \cdot \ldots \cdot 102}{4 \cdot 5 \cdot 6 \cdot \ldots \cdot 101} = \frac{2}{100} \cdot \frac{102}{4} = 0.51.$$

□

Problem 4.241. *Natural numbers from 1 to 2020 are written in a row: 123456789101112...20192020. How many times is number 12 repeated in this multi-digit number?*

(A) 155   (B) 105   (C) 165   (D) 150   (E) 175

Solution. Answer. (A)
Note that among three-digit numbers of the form $\overline{12a}$ number 12 is repeated 10 times and among the numbers of the form $\overline{a12}$ number 12 is repeated 9 times.
In a similar way, among the four-digit numbers of the forms $\overline{12ab}$, $\overline{a12b}$, $\overline{ab12}$ number 12 is repeated 100, 10 and 11 times, respectively.
Number 12 can be formed once by consecutive two-digit numbers of the forms $\overline{a1}$ and $\overline{2b}$.
Number 12 can be formed 10 times by the consecutive three-digit numbers of the forms $\overline{ab1}$ and $\overline{2cd}$.
Number 12 can be formed by four-digit numbers 2001, 2002, 2011, 2012.
Therefore, number 12 is repeated in this multi-digit number 155 times, as

$$1 + 10 + 9 + 100 + 10 + 11 + 1 + 1 + 10 + 2 = 155.$$

□

Problem 4.242. *Given a point M inside of the right triangle with legs AC and BC. Given that $\angle MAC = \angle MCA = 30°$ and $\dfrac{AC}{BC} = \dfrac{\sqrt{3}}{2}$. What is the value of the measure (in degrees) of $\angle AMB$?*

(A) 120 (B) 135 (C) 150 (D) 160 (E) 170

Solution. Answer. (C)
Let $MH \perp AC$ (see the figure).

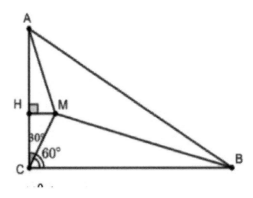

Triangle AMC is an isosceles triangle, therefore $AH = HC$.
We have that
$$\angle HMC = 60° = \angle MCB,$$
$$\dfrac{HM}{MC} = \dfrac{1}{2},$$
$$\dfrac{MC}{BC} = \dfrac{HC \cdot \dfrac{2}{\sqrt{3}}}{BC} = \dfrac{\dfrac{1}{2}AC \cdot \dfrac{2}{\sqrt{3}}}{BC} = \dfrac{1}{2}.$$

Thus, it follows that
$$\dfrac{HM}{MC} = \dfrac{MC}{BC}.$$

We obtain that triangles HCM and MBC are similar, therefore
$$\angle CMB = \angle MHC = 90°.$$

We have that
$$\angle AMB = 360° - \angle AMC - \angle CMB = 360° - 120° - 90° = 150°.$$

□

Problem 4.243. *Let u and v be positive numbers, such that $|u - v| \geq 1$. What is the smallest possible value of the expression $uv + \dfrac{u}{v} + \dfrac{v}{u}$?*

(A) 3 (B) 3.5 (C) 4 (D) $\sqrt{35}$ (E) 6

Solution. Answer. (C)
Note that
$$uv + \dfrac{u}{v} + \dfrac{v}{u} = uv + \dfrac{(u-v)^2}{uv} + 2 \geq uv + \dfrac{1}{uv} + 2 =$$
$$= \left(\sqrt{uv} - \dfrac{1}{\sqrt{uv}}\right)^2 + 4 \geq 4.$$

Thus, it follows that
$$uv + \frac{u}{v} + \frac{v}{u} \geq 4.$$
Note that, if
$$u = \frac{\sqrt{5}+1}{2},$$
and
$$v = \frac{\sqrt{5}-1}{2},$$
then $u - v = 1$ and $uv = 1$. Therefore
$$uv + \frac{u}{v} + \frac{v}{u} = uv + \frac{(u-v)^2}{uv} + 2 = 4.$$

Thus, the smallest possible value of the given expression is 4. $\square$

Problem 4.244. *What is the total number of all three-digit numbers, not containing any zero digit, for which there is a digit such that after erasing that digit the obtained two-digit number is divisible by 3? For example three-digit numbers 121 and 123 satisfy these conditions.*

(A) 120   (B) 159   (C) 729   (D) 540   (E) 513

Solution. Answer. (E)
A three-digit number that does not have any zero digit and does not satisfy the assumptions of the problem, we call an "insignificant" number.
Note that the total number of "insignificant" numbers, such that the sum of the digits for each of them is divisible by 3 is equal to
$$2 \cdot 3 \cdot 3 \cdot 3 = 54.$$
The total number of "insignificant" numbers, such that the sum of all digits for each of them leaves a remainder of 1 after division by 3 is equal to
$$3 \cdot 3 \cdot 3 \cdot 3 = 81.$$
The total number of "insignificant" numbers, such that the sum of all digits for each of them leaves a remainder of 2 after division by 3 is equal to
$$3 \cdot 3 \cdot 3 \cdot 3 = 81.$$
Therefore, the answer is
$$9 \cdot 9 \cdot 9 - 54 - 2 \cdot 81 = 513.$$
$\square$

Problem 4.245. *Let $ABCD$ be a rhombus, such that $\angle A = 45°$ and $AC = 13$. Assume that ray BD intersects the circumcircle of triangle ABC at point E. What is the value of the length of line segment DE?*

(A) 12   (B) 13   (C) $13\sqrt{2}$ (D) 20   (E) 26

Solution. Answer. (B)
Note that line BD is the perpendicular bisector of line segment AC, hence lige segment BE is the diameter of circumcircle of triangle ABC (see the figure).

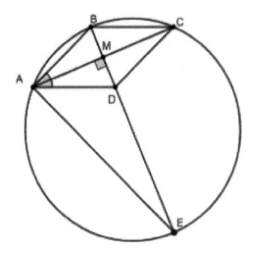

Thus, it follows that $\angle BAE = 90°$.

From triangle ABC according to law of sines, we obtain that

$$\frac{AC}{\sin 135°} = 2R = BE.$$

Hence, we deduce that $AC = \sqrt{2}R$.

Let $x = BM$. From triangle ABE we obtain that

$$\left(\frac{\sqrt{2}R}{2}\right)^2 = x(2R - x).$$

Thus, it follows that

$$x = R - \frac{\sqrt{2}R}{2},$$

and

$$DE = 2R - 2x = \sqrt{2}R = AC = 13.$$

$\square$

Problem 4.246. *Two-digit number is called "beautiful", if it does not end with 0 and is divisible by the sum of its digits. What is the value of the sum of all "beautiful" numbers?*

(A) 627   (B) 507   (C) 417   (D) 330   (E) 210

Solution. Answer. (A)

Let $\overline{xy}$ be a "beautiful" number and d be the greatest common divisor of x and y. Then $x = dx_1$ and $y = dy_1$, where x_1, y_1 are positive integers.

According to the assumption of the problem, we have that

$$(x + y) \mid \overline{xy}.$$

We have that

$$\overline{xy} = (x + y) + 9x.$$

Thus, it follows that

$$(x + y) \mid 9x.$$

We have that

$$(x_1 + y_1) \mid 9.$$

Therefore $x_1 + y_1 = 3$ or $x_1 + y_1 = 9$ (see the chart).

x_1	y_1	$\overline{xy}$
1	2	12, 24, 36, 48
2	1	21, 42, 63, 84
1	8	18
2	7	27
4	5	45
5	4	54
7	2	72
8	1	81

Thus, the sum of all "beautiful" numbers is:
$$12 \cdot 10 + 21 \cdot 10 + 9(2 + 3 + 5 + 6 + 8 + 9) = 627.$$

□

Problem 4.247. *An ant moves from the bottom left corner to the top right corner of 4×8 rectangular grid. It can move only on the sides of unit cells, such that one move is either going up 1 unit or going down 1 unit or going right 1 unit. Given that the ant cannot pass twice the same side of any unit cell. What is the probability of the event that the ant passes through the center of symmetry of given 4×8 rectangular grid?*

(A) $\dfrac{1}{2}$ (B) $\dfrac{8}{25}$ (C) $\dfrac{2}{3}$ (D) $\dfrac{17}{25}$ (E) $\dfrac{16}{25}$

Solution. Answer. (D)
Let us put into correspondance with each route of the ant an eight-digit number consisting only of the digits 1, 2, 3, 4, 5 (see the figure).

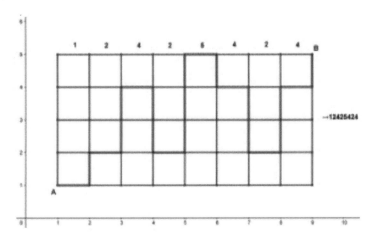

Therefore, the number of all possible routes is equal to 5^8.
If the route does not pass through the center of the symmetry, then for the eight-digit number $\overline{abcdefgh}$ corresponding to that route we have that $d, e \in \{1, 2\}$ or $d, e \in \{4, 5\}$.
Thus, the probability of an event that the route does not pass through the center of the symmetry is:
$$\frac{5^6 \cdot 8}{5^8} = \frac{8}{25}.$$

Therefore, the probability of the event that the ant passes through the center of symmetry of given 4×8 rectangular grid is equal to $\frac{17}{25}$, as
$$1 - \frac{8}{25} = \frac{17}{25}.$$

□

Problem 4.248. *8 soccer teams participated in a tournament. Each two teams played with each other only once (a winning team gets 3 points, a losing team gets 0 points, while a draw is 1 point for each team). Given that in the end the difference of the sum of the points of the teams at the first four places and the sum of the points of the teams at the last four places is equal to 54. How many points does the team at the last place has?*

(A) 0   (B) 1   (C) 2   (D) 3   (E) 10

Solution. Answer. (D)
Let m be the sum of the points of the teams at the first 4 places and n be the sum of the points of the teams at the last 4 places. Note that
$$m \leq 3 \cdot 4 \cdot 4 + 3 \cdot \binom{4}{2},$$
and
$$n \geq 2 \cdot \binom{4}{2}.$$

Thus, it follows that $m \leq 66$ and $n \geq 12$.
From the last two inequalities, we deduce that
$$m - n \leq 54.$$

Therefore $m = 66$ and $n = 12$.
We have obtained that the teams at the first 4 places have finished their games together with a draw and have lost to other teams.
Therefore, the team at the last place has made 3 draws and lost all other games.
Thus, it follows that the team at the last place has 3 points. □

Problem 4.249. *8 rectangular prisms with sizes $1 \times 1 \times 2$ must be placed in a rectangular prism box with size $2 \times 2 \times 4$. Given that all faces of the rectangular prism box are colored in different colors. In how many ways is it possible to do that?*

(A) 28   (B) 40   (C) 64   (D) 80   (E) 100

Solution. Answer. (E)
Assume that $2n$ rectangular prism boxes with sizes $1 \times 1 \times 2$ can be placed in a rectangular prism box with sizes $2 \times 2 \times n$ in x_n different ways.
And $2n - 1$ rectangular prism-boxes with sizes $1 \times 1 \times 2$ can be placed in y_n different ways into the box shown in the figure.

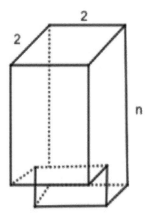

Note that
$$x_1 = 2, y_1 = 1.$$
Note also that
$$x_{k+1} = 2x_k + 4y_k,$$
and
$$y_{k+1} = x_k + y_k.$$
Thus, it follows that
$$y_2 = 3, x_2 = 2 \cdot 2 + 4 \cdot 1 = 8.$$
$$y_3 = 11, x_3 = 2 \cdot 8 + 4 \cdot 3 = 28,$$
$$x_4 = 2 \cdot 28 + 4 \cdot 11 = 100.$$

Therefore, it is possible to do in 4 different ways. □

Problem 4.250. *At least how many numbers should be erased from the list of numbers* $1, \frac{1}{2}, \frac{1}{3}, ..., \frac{1}{30}$ *in order to be able to split the rest of the numbers into two groups, such that the sum of all elements of each group are equal to each other?*

(A) 8   (B) 11   (C) 12   (D) 14   (E) 16

Solution. Answer. (B)

At first, let us formulate the following lemma.

Lemma. Let $m_1, m_2, ..., m_k$ be integers, $n_1, n_2, ..., n_k$ be positive integers, such that each of the fractions $\frac{m_1}{n_1}, \frac{m_2}{n_2}, ..., \frac{m_k}{n_k}$ is irreducible. Let p be a prime number and α be a positive integer, such that $p^\alpha \mid n_1$, but n_2 is not divisible by p^α,..., n_k is not divisible by p^α. Then, it follows that

$$\frac{m_1}{n_1} + \frac{m_2}{n_2} + ... + \frac{m_k}{n_k} \neq 0.$$

Proof of the lemma. Note that, the least common multiple of $n_1, n_2, ..., n_k$ is a number of the form $p^\alpha \cdot n$, where n is not divisible by p. Therefore

$$\frac{m_1}{n_1} + \frac{m_2}{n_2} + ... + \frac{m_k}{n_k} = \frac{m + pa}{p^\alpha \cdot n},$$

where m is not divisible by p, and m, a are integers.
Thus it follows that
$$\frac{m_1}{n_1} + \frac{m_2}{n_2} + ... + \frac{m_k}{n_k} \neq 0.$$

This ends the proof of the lemma.

According to the lemma, we have that each of the following numbers should be erased:
$$\frac{1}{29}, \frac{1}{27}, \frac{1}{25}, \frac{1}{23}, \frac{1}{19}, \frac{1}{17}, \frac{1}{16}.$$

Note that
$$\frac{1}{11} + \frac{1}{22} = \frac{3}{22},$$
and
$$\frac{1}{11} - \frac{1}{22} = \frac{1}{22},$$
therefore according to the lemma each of the numbers $\frac{1}{11}, \frac{1}{22}$ should be erased.

In a similar way, each of the numbers $\frac{1}{13}, \frac{1}{26}$ should be erased.

We have obtained that at least 11 numbers should be erased.

Note also that
$$\left(\frac{1}{7} + \frac{1}{21}\right) + \left(\frac{1}{5} + \frac{1}{10} + \frac{1}{20}\right) + \left(\frac{1}{8} + \frac{1}{24}\right) + \left(\frac{1}{2} + \frac{1}{3} + \frac{1}{6}\right) =$$
$$= \left(\frac{1}{12} + \frac{1}{14} + \frac{1}{28}\right) + \left(\frac{1}{4} + \frac{1}{15} + \frac{1}{30}\right) + \left(\frac{1}{9} + \frac{1}{18}\right) + 1.$$

□

4.11 Solutions of AMC 10 type practice test 11

Problem 4.251. *What is the value of the expression* $\dfrac{2020! + 2018!}{2017! \cdot (2019^3 - 1)}$?

(A) 2020 (B) 1 (C) $\dfrac{1}{2017}$ (D) $\dfrac{1}{2018}$ (E) $\dfrac{1}{2020}$

Solution. Answer. (B)
Note that
$$\frac{2020! + 2018!}{2017! \cdot (2019^3 - 1)} = \frac{2018!(2019 \cdot 2020 + 1)}{2017! \cdot (2019^3 - 1^3)} =$$
$$= 2018 \cdot \frac{2019^2 + 2019 + 1}{(2019 - 1)(2019^2 + 2019 + 1)} = 1.$$

□

Problem 4.252. *Which of the following statements is true? The sum of four consecutive integers can be:*

(A) 0 (B) Odd number (C) Multiple of 4 (D) A square of an integer (E) Even number

Solution. Answer. (E)
Consider four consecutive integers $n - 1, n, n + 1, n + 2$.
Note that, their sum is
$$4n + 2 = 2(2n + 1).$$
The sum of four consecutive integers can not be 0, an odd number, multiple of 4, or a square of an integer. On the other hand, we have that $2(2n + 1)$ is an even number.
Thus, it follows that the sum of four consecutive integers is even. □

Problem 4.253. *The median of numbers $1, 2, x, 13$ is equal to 3. What is the value of the mean of these numbers?*

(A) 5 (B) 6 (C) 10 (D) 11 (E) 12

Solution. Answer. (A)
Let us prove that
$$1 < x < 13.$$
We proceed the proof by contradiction argument.
If $x \leq 1$, then according to the assumption of the problem we have that $\dfrac{1+2}{2} = 3$. This leads to a contradiction. In a similar way, the case $x \geq 13$ leads to a contradiction.
Hence, we obtain that
$$\frac{2 + x}{2} = 3.$$
Thus, it follows that
$$\frac{1 + 2 + x + 13}{4} = \frac{1 + 6 + 13}{4} = 5.$$

□

Problem 4.254. *Let a and b be integers, such that $a \cdot b = 2020$. What is the total number of all such pairs (a,b)?*

(A) 10 (B) 11 (C) 12 (D) 24 (E) 30

Solution. Answer. (D)
Note that the total number of pairs (a,b) is equal to the total number of integer divisors of 2020, where a and b are integers and $a \cdot b = 2020$.
On the other hand, this number is equal to twice the number of all positive integer divisors of 2020.
We have that
$$2020 = 2^2 \cdot 5 \cdot 101.$$
Thus, it follows that the total number of all such pairs (a,b) is:
$$2 \cdot (2+1)(1+1)(1+1) = 24.$$

$\square$

Problem 4.255. *23 students sit in three rows in the classroom (there is at least one student in each row). Given that 20%, 25% and 10% of the students sitting respectively in the first, second and third row play basketball. In total, how many students play basketball?*

(A) 3 (B) 4 (C) 5 (D) 6 (E) 7

Solution. Answer. (B)
Let the number of students sitting in the first row be m, the number of students sitting in the second row be n and the number of students sitting in the third row be p.
Therefore, the number of students sitting in the first row and who play basketball is equal to $\dfrac{m \cdot 20}{100}$, the number of students sitting in the second row and who play basketball is equal to $\dfrac{n \cdot 25}{100}$, the number of students sitting in the third row and who play basketball is equal to $\dfrac{p \cdot 10}{100}$.
Thus, it follows that $5 \mid m$, $4 \mid n$ and $10 \mid p$.
Hence, we obtain that
$$m \geq 5, n \geq 4,$$
and
$$m + n \geq 9.$$
Therefore $p \leq 14$.
Taking all above mentioned into consideration, one can easily deduce that
$$p = 10, m = 5, n = 8.$$
Thus, it follows that in total 4 students play basketball, as
$$\frac{m}{5} + \frac{n}{4} + \frac{p}{10} = 4.$$

$\square$

Problem 4.256. *Let ABCD be a square and AMNK be a rectangle, such that their perimeters are equal (see the figure). Which of the following statements holds true?*

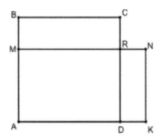

(A) $RC < RN$ (B) $RC > RN$ (C) $Area(MBCR) < Area(DRNK)$ (D) $Area(ABCD) = Area(AMNK)$ (E) $Area(ABCD) > Area(AMNK)$

Solution. Answer. (E)
Let us remove the segments with the same lengths from square $ABCD$ and rectangle $AMNK$ (see the figure).

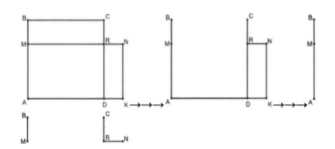 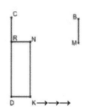 

According to the assumption of the problem, we have that
$$2RC = 2RN.$$
Thus, it follows that
$$RC = RN.$$
Therefore, statements (A) and (B) do not hold true.
We have that $RC = RN$ and $BC = CD > NK$, hence
$$Area(MBCR) > Area(DRNK).$$
Therefore, statements (C) and (D) do not hold true, statement (E) holds true. □

Problem 4.257. *Four numbers are chosen out of five consecutive integers. The sum of chosen numbers is equal to 2020. What is the value of the sum of these five consecutive integers?*

(A) 1000   (B) 2000   (C) 2500   (D) 2525   (E) 3225

Solution. Answer. (D)
Let these five consecutive integers be $n - 2, n - 1, n, n + 1, n + 2$.
According to the assumption of the problem, we have that
$$n - 2 + n - 1 + n + n + 1 + n + 2 - (n + r) = 2020,$$

where $r \in \{-2, -1, 0, 1, 2\}$.
Note that, from the equation
$$4n - r = 2020,$$
it follows that $4 \mid r$. Therefore $r = 0$ and $n = 505$.
Note that the sum of these five numbers is equal to $5n$.
Thus, it follows that the sum of these five numbers is equal to 2525. $\square$

Problem 4.258. *A paper rectangle is cut into three rectangles. The sum of the perimeters of those three rectangles is 32 and the perimeter of the initial paper rectangle is 16. What is the value of the area of the initial paper rectangle?*

(A) 7   (B) 12   (C) 15   (D) 15.5   (E) 16

Solution. Answer. (E)
Let us denote the lengths of the initial paper rectangle by a and b.
Note that, the sum of the perimeters of these 3 rectangles is either less than 32 (pic. 1) or equal to $6a + 2b$ (pic. 2).

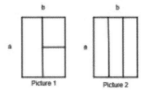

Thus, it follows that
$$a + b = 8,$$
and
$$3a + b = 16.$$
Hence, we obtain that
$$a = 4, b = 4.$$
Therefore, the area of the initial paper rectangle is equal to 16. $\square$

Problem 4.259. *Three sisters together bought three bracelets of different colors. In how many different ways can they wear all these three bracelets?*

(A) 27   (B) 48   (C) 216   (D) 264   (E) 336

Solution. Answer. (E)
Let us perform the following casework.
The number of ways when one of the sisters wears all three bracelets and the two other sisters do not wear any bracelet is equal to:
$$3 \cdot 2 \cdot 3! + 3 \cdot 3 \cdot 2 \cdot 2 = 72.$$

The number of ways when one of the sisters wears two bracelets, the other sister wears one bracelet and the third does not wear any bracelet is equal to:
$$6 \cdot 2 \cdot 2 \cdot 3 \cdot 3 = 216.$$

The number of ways when each sister wears one bracelet is equal to
$$3 \cdot 2 \cdot 1 \cdot 2 \cdot 2 \cdot 2 = 48.$$

Therefore, they can wear all these three bracelets in $72 + 216 + 48 = 336$ different ways. $\square$

Problem 4.260. *Let BE be a median of triangle ABC and CF be a median of triangle BEC. Given that $AF = AE$ and $CF = 20$. What is the length of line segment AB?*

(A) 16   (B) 17   (C) 18   (D) 19   (E) 20

Solution. Answer. (E)
According to the assumption of the problem, we have that AEF is an isosceles triangle (see the figure).

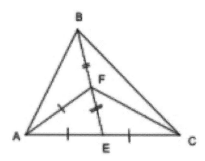

Thus, it follows that
$$\angle AEF = \angle AFE.$$
We obtain that
$$\angle AFB = 180° - \angle AFE = 180° - \angle AEF = \angle CEF.$$
We deduce that
$$AF = CE, BF = EF,$$
and
$$\angle AFB = \angle CEF.$$
Note that triangles AFB and CEF are congruent triangles. Thus, it follows that
$$AB = CF = 20.$$

□

Problem 4.261. *The factory did not work on every Saturday and Sunday of February. On the nth day of each week it produced $(6-n)^2$ devices, where $n \in \{1,2,3,4,5\}$. Given that in total the factory produced 236 devices during the month of February. Which day of the week was the last day of that February?*

(A) Monday   (B) Tuesday   (C) Wednesday   (D) Thursday   (E) Friday

Solution. Answer. (B)
Note that during the first 28 days of February the factory produced $4(5^2 + 4^2 + 3^2 + 2^2 + 1^2) = 220$ devices.
Therefore, that February had 29 days and on the last day the factory produced $236 - 220 = 16$ devices.
Hence, the last day of that February was Tuesday.   □

Problem 4.262. *The probability that it will not rain for two consecutive days is $\frac{1}{6}$, the probability that it will rain on both days is $\frac{1}{3}$. The probability that it will rain only on one of those two days is $\frac{1}{2}$. What is the probability that it will rain exactly two days out of six consecutive days?*

(A) $\frac{11}{72}$ (B) $\frac{5}{18}$ (C) $\frac{1}{4}$ (D) $\frac{1}{2}$ (E) $\frac{2}{3}$

Solution. Answer. (A)

Let us divide these six consecutive days into three pairs of two consecutive days.

Then, the probability that it will rain exactly two days out of six consecutive days is equal to:

$$\frac{1}{3} \cdot \frac{1}{6} \cdot \frac{1}{6} + \frac{1}{6} \cdot \frac{1}{3} \cdot \frac{1}{6} + \frac{1}{6} \cdot \frac{1}{6} \cdot \frac{1}{3} + \frac{1}{2} \cdot \frac{1}{2} \cdot \frac{1}{6} +$$

$$\frac{1}{2} \cdot \frac{1}{6} \cdot \frac{1}{2} + \frac{1}{6} \cdot \frac{1}{2} \cdot \frac{1}{2} = \frac{11}{72}.$$

$\square$

Problem 4.263. *Positive integer n is called "amazing", if the sum of its two greatest divisors is equal to 42. What is the total number of all "amazing" numbers?*

(A) 4   (B) 3   (C) 5   (D) 2   (E) 6

Solution. Answer. (B)

Assume the smallest prime divisor of the "amazing" number n is p.

Note that, in this case the two greatest divisors of n are n and $\frac{n}{p}$.

According to the assumption of the problem, we have that the sum of two greatest divisors of n is equal to 42, therefore

$$n + \frac{n}{p} = 42.$$

Thus, it follows that

$$n = \frac{42p}{p+1}.$$

We deduce that

$$(p+1) \mid 42.$$

Taking this into consideration and as p is prime, we obtain that

$$p \in \{2, 5, 13, 41\}.$$

A straightforward verification shows that only the values $p = 2$, $p = 5$, $p = 41$ satisfy the assumptions of the problem and in these cases $n = 28$, $n = 35$, $n = 41$, respectively. Therefore, the total number of all "amazing" numbers is equal to 3.

$\square$

Problem 4.264. *Let $ABCDEF$ be an inscribed hexagon. Given that the measures (in degrees) of angles A, B, C, D, E form an increasing arithmetic sequence (in this order) with a difference of $10°$. What is the measure (in degrees) of angle A?*

(A) 90   (B) 100   (C) 120   (D) 135   (E) 150

Solution. Answer. (B)

Let $\angle A = \alpha$. According to the assumption of the problem, we have that the measures (in degrees) of angles A, B, C, D, E form an arithmetic sequence with a difference of $10°$. Therefore

$$\angle B = \alpha + 10°, \angle C = \alpha + 20°, \angle D = \alpha + 30°, \angle E = \alpha + 40°.$$

Note that (see the figure)

$$\angle A + \angle C + \angle E = 180° - \angle BDF + 180° - \angle BFD + 180° - \angle FBD = 540° - 180° = 360°.$$

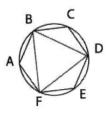

Hence, we obtain that
$$\alpha + \alpha + 20° + \alpha + 40° = 360°.$$
Thus, it follows that
$$\alpha = 100°.$$

□

Problem 4.265. *At most how many three-digit numbers can have the same sum of the digits?*

(A) 62 (B) 69 (C) 70 (D) 80 (E) 120

Solution. Answer. (C)
Let us consider the following table (see the figure).

x+y	0	1	2	3	4	5	6	7	8	9	10	11	12	13	14	15	16	17	18
The number of pairs of (x, y) digits	1	2	3	4	5	6	7	8	9	10	9	8	7	6	5	4	3	2	1

Taking this table into consideration, we have that the greatest number of three-digit numbers $\overline{zxy}$ with equal sums of the digits is equal to the sum of the numbers written in 9 consecutive cells of the table the sum of which is the greatest.
Thus, it follows that the greatest number of all such three-digit numbers is:
$$10 + 2 \cdot 9 + 2 \cdot 8 + 2 \cdot 7 + 2 \cdot 6 = 70.$$

Therefore, the total number of all such three-digit numbers is equal to 70 (and the sum of the digits for each of them is equal to 14). □

Problem 4.266. *What is the value of the sum of the digits of the smallest positive integer n, where $n \neq 1$, such that n and n^3 leave the same remainder after division by 2020^2?*

(A) 2 (B) 3 (C) 4 (D) 6 (E) 13

Solution. Answer. (C)
According to the assumption of the problem, we have that
$$2020^2 \mid (n^3 - n).$$
Note that
$$(n^3 - n) = (n-1)n(n+1),$$
and
$$2020^2 = 2^4 \cdot 5^2 \cdot 101^2.$$

Note also that the difference of any two of the numbers $n-1$, n, $n+1$ is not greater than 2. Therefore, the greatest common divisor of any two of these numbers is also not greater than 2. Hence, we obtain that either
$$101^2 \mid (n-1),$$

or
$$101^2 \mid n,$$
or
$$101^2 \mid (n+1).$$

We deduce that
$$n \geq 101^2 - 1 = 10200.$$

Note that, if $n = 10200$, then
$$(n-1)n(n+1) = (101^2 - 2)(101^2 - 1) \cdot 101^2$$
is not divisible by 2020^2 and
$$2020^2 \mid (101^2 - 1) \cdot 101^2 \cdot (101^2 + 1).$$

Thus, it follows that
$$n = 101^2 = 10201.$$

Therefore, the sum of the digits of the smallest possible such n is equal to 4. □

Problem 4.267. *Let AD be the bisector of angle BAC in triangle ABC. Given that $BD = 3$, $CD = 5$ and $AC - AB = 4$. What is the measure (in degrees) of $\angle ABC$?*

(A) 120   (B) 100   (C) 90   (D) 75   (E) 60

Solution. Answer. (C)
Let us choose point E on side AC, such that $AE = AB$ (see the figure).

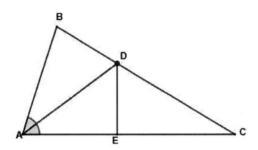

According to the assumption of the problem, we have that
$$EC = AC - AE = AC - AB = 4.$$

Note that triangles ABD and AED are congruent triangles. Therefore
$$BD = DE, \angle ABD = \angle AED.$$

Thus, it follows that
$$DE = 3, EC = 4, CD = 5.$$

Hence, according to the converse theorem of Pythagorean theorem from triangle DEC we obtain that
$$\angle DEC = 90°.$$

We deduce that
$$\angle ABC = \angle ABD = \angle AED = 180° - \angle DEC = 90°.$$
□

Problem 4.268. *What is the total number of all five-digit numbers divisible by 13 and ending with 12?*

(A) 23   (B) 27   (C) 60   (D) 68   (E) 69

Solution. Answer. (E)
Let $\overline{abc12}$ be a five-digit number divisible by 13. Note that
$$\overline{abc12} = \overline{abc00} + 12 = 100 \cdot \overline{abc} + 12 =$$
$$= 104 \cdot \overline{abc} - 4(\overline{abc} - 3) = 13 \cdot 8 \cdot \overline{abc} - 4(\overline{abc} - 3).$$
Thus, it follows that $13 \mid \overline{abc12}$ if and only if $13 \mid (\overline{abc} - 3)$.
Now, we need to find the total number of all three-digit numbers which leave a remainder of 3 after division by 13.
Such three-digit numbers form an arithmetic sequence with the first term equal to 107 and with a common difference of 13.
Taking this into consideration, we obtain that
$$107 + 13(n-1) \leq 999.$$
Hence, we deduce that
$$n \in \{1, 2, ..., 69\}.$$
Therefore, the total number of all such five-digit numbers is equal to 69. $\square$

Problem 4.269. *Let E and F be respectively the midpoints of sides AB and BC of square $ABCD$. Line segment EC intersects line segments AF, DF at points P, K respectively. Given that $AB = 5\sqrt{6}$. What is the area of triangle FPK?*

(A) $\sqrt{6}$ (B) 3   (C) 4   (D) $2\sqrt{6}$ (E) 5

Solution. Answer. (E)
Note that triangles EBC and FCD are congruent triangles (see the figure).

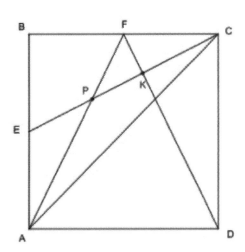

Thus, it follows that
$$\angle BEC = \angle CFD.$$
Given that
$$\angle FKC = 180° - \angle CFK - \angle KCF = 180° - \angle CFD - \angle ECB =$$
$$= 180° - \angle BEC - \angle ECB = 90°.$$

Hence, we obtain that
$$\angle FKC = 90°.$$

Note that AF and CE are medians in triangle ABC, therefore
$$Area(PFC) = \frac{1}{6}Area(ABC) = \frac{1}{12}Area(ABCD) = 12.5.$$

In the right triangle FCD, we have that
$$\frac{FK}{KD} = \frac{FK \cdot FD}{KD \cdot FD} = \frac{CF^2}{CD^2} = \frac{1}{4}.$$

Thus, it follows that
$$Area(CKF) = \frac{1}{5}Area(FCD) = \frac{1}{20}Area(ABCD) = 7.5.$$

Hence, we deduce that
$$Area(FPK) = Area(PFC) - Area(CKF) = 12.5 - 7.5 = 5.$$

□

Problem 4.270. *What is the total number of all positive divisors of the number 10! which are multiples of 3?*

(A) 100   (B) 200   (C) 216   (D) 300   (E) 420

Solution. Answer. (C)
We have that
$$10! = 2^8 \cdot 3^4 \cdot 5^2 \cdot 7.$$
Therefore, a positive divisors multiples of 3 of the number 10! have the form $2^\alpha \cdot 3^\beta \cdot 5^\gamma \cdot 7^\theta$, where $\alpha \in \{0, 1, ..., 8\}$, $\beta \in \{1, 2, 3, 4\}$, $\gamma \in \{0, 1, 2\}$, $\theta \in \{0, 1\}$.
Thus, it follows that the total number of all positive divisors of the number 10! which are multiples of 3 is equal to the total number of quadruples $(\alpha, \beta, \gamma, \theta)$, that is $9 \cdot 4 \cdot 3 \cdot 2 = 216$. □

Problem 4.271. *A three-digit number $\overline{abc}$ written using some of the digits 0, 1, 2, 3, 4 is randomly chosen. What is the probability that the inequalities $|a - b| \geq 2$ and $|b - c| \geq 2$ simultaneously hold true?*

(A) 0.05   (B) 0.1   (C) 0.2   (D) 0.23   (E) 0.25

Solution. Answer. (D)
Let $i \to j$ means that in three-digit number $\overline{abc}$ in the right side of the digit i is written digit j.
All three-digit numbers $\overline{abc}$ satisfying the inequalities $|a - b| \geq 2$ and $|b - c| \geq 2$ can be shown as follows (see the figure).

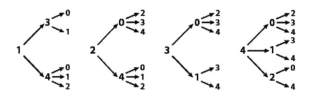

Therefore, the total number of all possible outcomes is equal to $4 \cdot 5 \cdot 5 = 100$ and the total number of all favorable outcomes is equal to 23.

Thus, the probability that the given inequalities simultaneously hold true is equal to $\frac{23}{100} = 0.23$. □

Problem 4.272. *At most how many pairwise distinct integers are there, such that the positive difference of any two of them is a prime number?*

(A) 3 (B) 4 (C) 5 (D) 6 (E) 10

Solution. Answer. (B)
Note that five such numbers do not exist. Otherwise according to the pigeonhole principle at least three of them have the same parity and the difference of the greatest and smallest of those numbers is an even number not less than 4, which cannot be a prime number.
An example of such four numbers is 2, 4, 7, 9. □

Problem 4.273. *Let integer a be the smallest value of variable x, such that the values of the functions $y = \dfrac{2x}{3} - \dfrac{1}{3}$ and $y = \dfrac{3x^2}{14} + \dfrac{x}{7} - \dfrac{1}{2}$ are positive integers. What is the value of the sum of the digits of the number a^2?*

(A) 4 (B) 5 (C) 6 (D) 7 (E) 10

Solution. Answer. (A)
According to the assumption of the problem, we have that
$$\frac{2a}{3} - \frac{1}{3} = m,$$
$$\frac{3a^2}{14} + \frac{a}{7} - \frac{1}{2} = n,$$
where m, n are positive integers and the value of m is the smallest possible.
Thus, it follows that
$$a = \frac{3m+1}{2},$$
$$\frac{3a^2}{14} + \frac{a}{7} - \frac{1}{2} = \frac{3(3m+1)^2}{56} + \frac{3m+1}{14} - \frac{1}{2} = \frac{27m^2 + 30m - 21}{56} = \frac{3(9m^2 + 10m - 7)}{56} = n.$$
Hence, we obtain that
$$m = 7, n = \frac{3 \cdot (63 + 10 - 1)}{8} = 27.$$
We deduce that
$$m = 7, a = 11.$$
Therefore $a^2 = 121$ and the sum of the digits of a^2 is equal to 4. □

Problem 4.274. Let $A_1B_1C_1$ be a triangle, such that $A_1B_1 = 4$, $A_1C_1 = 5$ and $B_1C_1 = 7$. Consider the sequence of triangles $A_nB_nC_1$ constructed as follows: for any positive integer n points B_{n+1} and A_{n+1} lie on sides C_1A_n and C_1B_n, respectively. Given that $\angle C_1B_{n+1}A_{n+1} = \angle C_1B_nA_n$ and that $A_nB_{n+1}A_{n+1}B_n$ is a tangential quadrilateral. What is the value of the sum of the perimeters of all triangles $A_nB_nC_1$?

(A) 18   (B) 20   (C) 24   (D) 32   (E) 60

Solution. Answer. (D)
Consider the following figure.

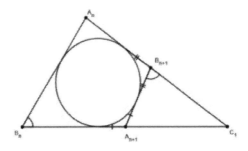

Let us denote:
$$C_1B_n = a_n, C_1A_n = b_n, A_nB_n = c_n.$$

We have that
$$a_1 = 7, b_1 = 5, c_1 = 4.$$

Note that
$$\angle C_1B_nA_n = \angle C_1B_{n-1}A_{n-1} = \ldots = \angle C_1B_1A_1.$$

Thus, it follows that triangles $A_nB_nC_1$ and $A_1B_1C_1$ are similar.
Let $\dfrac{c_n}{c_1} = \lambda_n$, then note that

$$a_{n+1} + b_{n+1} + c_{n+1} = 2 \cdot \frac{a_n + b_n - c_n}{2}.$$

Hence, we obtain that
$$\lambda_{n+1} \cdot a_1 + \lambda_{n+1} \cdot b_1 + \lambda_{n+1} \cdot c_1 = \lambda_n \cdot a_1 + \lambda_n \cdot b_1 - \lambda_n \cdot c_1.$$

Thus, we deduce that
$$\frac{\lambda_{n+1}}{\lambda_n} = \frac{a_1 + b_1 - c_1}{a_1 + b_1 + c_1} = \frac{1}{2}.$$

Therefore, the sum of the perimeters of all triangles $A_nB_nC_1$ is:
$$\frac{a_1 + b_1 + c_1}{1 - \dfrac{1}{2}} = 32.$$

Problem 4.275. *A permutation $a_1, a_2,..., a_8$ of the numbers 1, 2,..., 8 is called "charming" if for any number $i \in \{1, 2, ..., 8\}$ either $i \mid a_i$ or $\lfloor \frac{i}{2} \rfloor \mid a_i$. What is the total number of all "charming" permutations of numbers 1,2,...,8?*

(A) 252   (B) 216   (C) 192   (D) 120   (E) 96

Solution. Answer. (A)

Let the permutation $a_1, a_2,..., a_8$ of the numbers 1, 2,..., 8 be "charming".

According to the assumption of the problem, we have that each of the numbers $a_1, a_2,..., a_8$ can be equal to any of the numbers 1, 2,.., 8. Besides, either $2 \mid a_4$, $a_5 = 5$ or $2 \mid a_5$, $3 \mid a_6$, $a_7 = 7$ or $3 \mid a_7$, $4 \mid a_8$. Thus, it follows that either $a_8 = 4$ or $a_8 = 8$, $a_7 = 3$ or $a_7 = 6$ or $a_7 = 7$, $a_6 = 3$ or $a_6 = 6$, $a_5 = 5$ or a_5 is an even number, a_4 is an even number.

Let us consider the following cases.

Case 1. If $a_5 = 5$ and $a_7 = 7$.
Case 2. If $a_5 \neq 5$ and $a_7 = 7$.
Case 3. If $a_5 = 5$ and $a_7 \neq 7$.
Case 4. If $a_5 \neq 5$ and $a_7 \neq 7$.

Note that the number of "charming" permutations corresponding to case 1, case 2, case 3, case 4 is respectively equal to $(3 \cdot 3! + 2 \cdot 3!) \cdot 2 = 60$, $(3 \cdot 2 \cdot 3! + 2 \cdot 1 \cdot 3!) \cdot 2 = 96$, $2 \cdot 2 \cdot 2 \cdot 3 = 48$ and $2 \cdot 2 \cdot 2 \cdot 3 = 48$. Thus, the total number of all "charming" permutations of numbers 1,2,...,8 is $60 + 96 + 48 + 48 = 252$. □

4.12 Solutions of AMC 10 type practice test 12

Problem 4.276. *What is the value of the expression $(15^{-2} + 20^{-2} - 12^{-2})(15^2 + 20^2 - 12^2)$?*

(A) $\dfrac{1}{20}$ (B) $\dfrac{1}{12}$ (C) 1 (D) 0 (E) -1

Solution. Answer. (D)
Note that

$$15^{-2} + 20^{-2} - 12^{-2} = 5^{-2}(3^{-2} + 4^{-2}) - 12^{-2} = 5^{-2} \cdot 12^{-2}(4^2 + 3^2) - 12^{-2} = 12^{-2} - 12^{-2} = 0.$$

□

Problem 4.277. *At first, David and Anna ate 7 candies together, then David ate another 5 candies. In the end it turned out that David ate twice as many candies as Anna. How many candies did Anna eat?*

(A) 3 (B) 4 (C) 5 (D) 6 (E) 2

Solution. Answer. (B)
Note that after David ate another 5 candies, the number of candies they ate together became $7+5 = 12$.
According to the assumption of the problem, we have that Anna ate $\dfrac{1}{3}$ of these 12 candies.
Thus, it follows that Anna ate 4 candies. □

Problem 4.278. *Given two positive numbers, such that the first number is greater than the second number by 10%. Given also that the difference of these two numbers is equal to 3. What is the value of the first number?*

(A) 33 (B) 30 (C) 27 (D) 25 (E) 12

Solution. Answer. (A)
According to the assumption of the problem, we have that 10% of the second number is equal to 3. Thus, it follows that the second number is 30.
Taking into consideration that the first number is greater than the second number and that their difference is equal to 3, we obtain that the first number is equal to 33. □

Problem 4.279. *Five two-digit numbers are formed with the digits 0, 1, 2, 3, 4, 5, 6, 7, 8, 9, using each digit once. What is the possible smallest value of the sum of these five two-digit numbers?*

(A) 180 (B) 135 (C) 120 (D) 100 (E) 99

Solution. Answer. (A)
Let the sum of the first digits of those five two-digit numbers be x, therefore the sum of the second digits is $45 - x$.
Hence, the sum of these five two-digit numbers is $10x + 45 - x = 9x + 45$.
Therefore, the possible smallest value of the sum of these five two-digit numbers is $9 \cdot 15 + 45 = 180$. □

Problem 4.280. *Let a, b, c be real numbers, such that $ab - bc < 0$, $bc - ac < 0$ and $ab + ac < 0$. Which of the following conditions can hold true?*

(A) $a > 0, b > 0, c < 0$ (B) $a > 0, b > 0, c > 0$ (C) $a < 0, b < 0, c > 0$ (D) $a < 0, b > 0, c < 0$
(E) $a < 0, b > 0, c > 0$

Solution. Answer. (D)
Note that if the condition (A) holds true, then $ab - bc > 0$.
If the condition (B) holds true, then $ab + ac > 0$.
If the condition (C) holds true, then $ab - bc > 0$.
If the condition (E) holds true, then $bc - ac > 0$.
Note also that the condition (D) is possible, for example if $a = -2$, $b = 2$, $c = -1$. $\square$

Problem 4.281. *Given that today is not Wednesday. What is the probability of the event that tomorrow is Wednesday?*

(A) 0   (B) $\dfrac{1}{7}$ (C) $\dfrac{1}{6}$ (D) $\dfrac{2}{7}$ (E) $\dfrac{3}{7}$

Solution. Answer. (C)
Note that the number of all possible outcomes is 6 and the number of all favorable outcomes is 1. Therefore, the probability that tomorrow is Wednesday is $\dfrac{1}{6}$. $\square$

Problem 4.282. *The sum of the ages of all classmates is 120. In 6 years the sum of theirs ages will be twice the sum of their ages 3 years ago. What is the value of the arithmetic mean of their ages?*

(A) 8   (B) 10   (C) 12   (D) 15   (E) 16

Solution. Answer. (C)
Let n be the number of classmates. In 6 years the sum of all their ages will be $120 + 6n$.
On the other hand, 3 years ago the sum of their ages was $120 - 3n$.
According to the assumption of the problem, we have that
$$120 + 6n = 2(120 - 3n).$$
Thus, it follows that
$$12n = 120.$$
Therefore $n = 10$. $\square$

Problem 4.283. *Let $ABCD$ be a convex quadrilateral, such that $AD = 2BC$, $BD = BC$, $\angle ABC = 130°$ and $\angle BCD = 70°$. What is the measure (in degrees) of angle ADC?*

(A) 60   (B) 90   (C) 120   (D) 125   (E) 130

Solution. Answer. (E)
Consider the following figure.

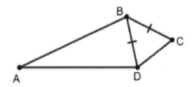

Given that $BD = BC$, therefore
$$\angle BDC = \angle BCD = 70°.$$

From triangle BDC, we obtain that
$$\angle DBC = 180° - 140° = 40°.$$
Thus, it follows that
$$\angle ABD = 130° - 40° = 90°.$$
Note that
$$AD = 2BC = 2BD,$$
and this implies that $\angle BAD = 30°$. Hence, we obtain that $\angle ADB = 60°$.
On the other hand, we have that
$$\angle ADC = \angle ADB + \angle BDC = 60° + 70° = 130°.$$

$\square$

Problem 4.284. *David solved 5 problems each Saturdays and Sundays, on each week-day he solved 6 problems. Given that in a few consecutive days David solved 70 problems. Which day of the week did he start to solve these problems?*

(A) Wednesday (B) Thursday (C) Friday (D) Sunday (E) Monday

Solution. Answer. (E)
Note that
$$6 \mid (70 - 2 \cdot 5),$$
and
$$6 \mid (70 - 8 \cdot 5).$$
Besides, if we have 8 Saturdays and Sundays, then the number of other days could not be 5.
Thus, the number of Saturdays and Sundays is 2 and the number of other days is 10.
Therefore, David started to solve these problems on Monday. $\square$

Problem 4.285. *Let $ABCD$ be a square with a side length $\sqrt{3}$. Let square $ABCD$ be rotated around point A by $30°$ and as a result we obtained square $AB'C'D'$. What is the area of the part that squares $AB'C'D'$ and $ABCD$ have in common?*

(A) 1 (B) 1.5 (C) $\sqrt{3}$ (D) 2 (E) 2.1

Solution. Answer. (C)
Consider the following figure.

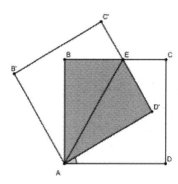

Note that right triangles ABE and AED' are congruent, therefore
$$\angle BAE = \angle D'AE = \frac{60°}{2} = 30°.$$

Hence, from triangle ABE we obtain that $BE = 1$ and
$$Area(ABE) = \frac{\sqrt{3}}{2}.$$
Thus, it follows that
$$Area(ABED') = 2Area(ABE) = \sqrt{3}.$$
□

Problem 4.286. *Given that the intersection points of the graphs of the functions $y = x^2 + ax + 1$ and $y = -x^2 - 3x + b$ are symmetric with respect to point $(-1, 5)$. What is the value of $a + b$?*

(A) 10   (B) 8   (C) 7   (D) 5   (E) -6

Solution. Answer. (B)
Let the graphs of the functions $y = x^2 + ax + 1$ and $y = -x^2 - 3x + b$ intersect at points (x_1, y_1) and (x_2, y_2).
Thus, it follows that x_1 and x_2 are the solutions of the following equation
$$x^2 + ax + 1 = -x^2 - 3x + b,$$
or equivalently
$$2x^2 + (a+3)x + 1 - b = 0.$$
According to Vieta's formula, we have that
$$x_1 + x_2 = -\frac{a+3}{2}.$$
According to the assumption of the problem, we have that
$$\frac{x_1 + x_2}{2} = -1, \frac{y_1 + y_2}{2} = 5.$$
Thus, it follows that
$$-\frac{a+3}{2} = -2.$$
Therefore $a = 1$. On the other hand, we have that
$$\frac{y_1 + y_2}{2} = 5.$$
We deduce that
$$x_1^2 + x_1 + 1 + x_2^2 + x_2 + 1 = 10,$$
$$(x_1 + x_2)^2 - 2x_1x_2 + x_1 + x_2 = 8.$$
Therefore $x_1 x_2 = -3$. Thus, it follows that
$$\frac{1-b}{2} = -3.$$
We obtain that $b = 7$. Hence
$$a + b = 7 + 1 = 8.$$
□

Problem 4.287. *Given three pairwise different increasing arithmetic sequences, each consisting of one hundred integer terms. Let n be the total number of all those terms, each of which simultaneously belongs to these three sequences. What is the greatest possible value of n?*

(A) 50   (B) 34   (C) 33   (D) 25   (E) 16

Solution. Answer. (B)
Note that for arithmetic sequences 0, 2, 4, 6,..., 198 and 0, 3, 6, 9,..., 297 and 0, 6, 12, 18,..., 594 we have that $n = 34$.
On the other hand, note that the case $n > 34$ is impossible.
As in this case for at least one of these arithmetic sequences, in between any two terms a and b which are simultaneously terms of the other two sequences, there will be at least two terms from that sequence.
Thus, it follows that the greatest possible value of n is equal to 34. □

Problem 4.288. *Three sisters bought 4 identical bracelets. In how many different ways can they wear those 4 bracelets? (They can wear each bracelet either on right or left hand).*

(A) 15   (B) 20   (C) 24   (D) 120   (E) 126

Solution. Answer. (E)
Assume the first sister wears x_1 bracelet(s) on her left hand and x_2 bracelet(s) on the right hand.
In the same way we define numbers x_3, x_4, x_5, x_6. We need to find the total number of non-negative integer solutions of the following equation

$$x_1 + x_2 + x_3 + x_4 + x_5 + x_6 = 4.$$

The total number of non-negative integer solutions of this equation is equal to

$$\binom{9}{5} = 126.$$

□

Problem 4.289. *What is the greatest possible number of all distinct prime numbers, such that the positive difference of any two of them is also a prime number?*

(A) 2   (B) 3   (C) 4   (D) 5   (E) 10

Solution. Answer. (B)
Note that if three of those prime numbers are odd, then the difference of the greatest and smallest of them is an even number greater than 2, which cannot be a prime number.
Thus, it follows that the total number of prime numbers satisfying the assumptions of the problem is not more than 3.
An example of three such prime numbers is 2, 5, 7.
Therefore, the greatest possible number of all distinct prime numbers satisfying the assumptions of the problem is equal to 3. □

Problem 4.290. *Let $\{a\} = a - \lfloor a \rfloor$, where $\lfloor a \rfloor$ is the greatest positive integer not greater than a. What is the total number of the solutions of the equation $x + \{x\}^2 = 2020$?*

(A) 0   (B) 1   (C) 2   (D) 3   (E) 2020

Solution. Answer. (C)
We have that

$$x = \lfloor x \rfloor + \{x\},$$

and
$$0 \leq x < 1.$$

Taking this and the assumption of the problem into consideration, it follows that we need to find the total number of the solutions of the following equation

$$\lfloor x \rfloor + \{x\} + \{x\}^2 = 2020.$$

Hence, we obtain that $\{x\} + \{x\}^2$ is an integer. We have that

$$0 \leq \{x\} + \{x\}^2 < 2.$$

Thus, it follows that either
$$\{x\} + \{x\}^2 = 0,$$
or
$$\{x\} + \{x\}^2 = 1,$$

where either $\{x\} = 0$ or $\{x\} = \dfrac{\sqrt{5}-1}{2}$.

Therefore, the solutions of the given equation are 2020 and $2019 + \dfrac{\sqrt{5}-1}{2}$.

Hence, we obtain that given equation has 2 solutions. $\square$

Problem 4.291. *Jack shoots at a moving circular disk that rotates around its center. The probability to score 1 point, 2 points, 3 points with one shot is $\dfrac{1}{2}, \dfrac{1}{3}, \dfrac{1}{6}$, respectively. What is the probability of the event that Jack scores 6 points with three shots?*

(A) $\dfrac{1}{7}$ (B) $\dfrac{11}{54}$ (C) $\dfrac{1}{6}$ (D) $\dfrac{1}{3}$ (E) $\dfrac{1}{2}$

Solution. Answer. (B)

Note that Jack can score 6 points in the following seven cases.

$$2 + 2 + 2 = 6.$$
$$1 + 2 + 3 = 6.$$
$$1 + 3 + 2 = 6.$$
$$2 + 1 + 3 = 6.$$
$$2 + 3 + 1 = 6.$$
$$3 + 1 + 2 = 6.$$
$$3 + 2 + 1 = 6.$$

Therefore, the probability of the event that Jack scores 6 points with three shots is

$$\frac{1}{3} \cdot \frac{1}{3} \cdot \frac{1}{3} + 6 \cdot \frac{1}{2} \cdot \frac{1}{3} \cdot \frac{1}{6} = \frac{11}{54}.$$

$\square$

Problem 4.292. *Given that the sum of three smallest positive divisors of a positive integer n is equal to 7 and the sum of three greatest divisors of n is equal to 84. What is the value of the sum of all digits of n?*

(A) 8   (B) 9   (C) 10   (D) 11   (E) 12

Solution. Answer. (E)

According to the assumption of the problem, we have that the sum of three smallest positive divisors of n is equal to 7. Therefore, three smallest positive divisors of n are 1, 2, 4.

In this case, three greatest divisors of n are n, $\frac{n}{2}$, $\frac{n}{4}$.

According to the assumption of the problem, we have the the sum of three greatest positive divisors of n is equal to 84. Thus, it follows that
$$n + \frac{n}{2} + \frac{n}{4} = 84.$$

Hence, we obtain that $n = 48$ and the sum of all digits of n is equal to 12. $\square$

Problem 4.293. *Let θ be the circumscribed circle of triangle ABC and $\angle A - \angle B = 50°$. Let the perpendicular bisector of line segment AB intersects the arc ACB of circle θ at point M. What is the measure (in degrees) of $\angle MAC$?*

(A) 20   (B) 25   (C) 30   (D) 15   (E) 10

Solution. Answer. (B)

We have that $MA = MB$ and $\angle MAC = \angle MBC$ (see the figure).

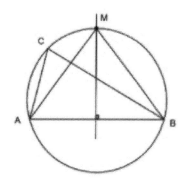

Let $\angle MAC = \alpha$, $\angle MAB = \beta$. Thus, it follows that
$$\angle A = \alpha + \beta,$$
$$\angle B = \beta - \alpha.$$

According to the assumption of the problem, we have that
$$\alpha + \beta - (\beta - \alpha) = 50°.$$

Hence, we obtain that $\alpha = 25°$.
Thus, it follows that $\angle MAC = 25°$. $\square$

Problem 4.294. *What is the value of the sum of all integers x, such that the the double inequality $|x+1| \leq |x-27| \leq |x+3|$ holds true?*

(A) 15 (B) 21 (C) 23 (D) 25 (E) 30

Solution. Answer. (D)
Note that all integer solutions of the inequality $|x+1| \leq |x-27|$ are all integers x, such that the distance from point $M(x)$ to point $A(-1)$ is not more than the distance from point $M(x)$ to point $B(27)$. Thus, it follows that
$$x \in \{13, 12, 11, ...\}.$$
In a similar way, from the condition $|x-27| \leq |x+3|$ we obtain that $x \in \{12, 13, ...\}$.
Therefore, the sum of all integers, such that given double inequality holds true is $12 + 13 = 25$. □

Problem 4.295. *Let $ABCD$ be a square with a side lenght of $2\sqrt{5}$. Let M, N, P, K be the midpoints of sides AB, BC, CD, AD, respectively. What is the radius of the circle that is tangent to each of line segments AN, DN, AP, BP, BK, CK, CM, DM?*

(A) $\dfrac{\sqrt{5}}{2}$ (B) 1 (C) $\dfrac{\sqrt{3}}{2}$ (D) $\dfrac{\sqrt{2}}{2}$ (E) $\dfrac{1}{2}$

Solution. Answer. (B)
Consider the following figure.

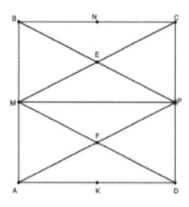

Note that we need to find the radius of the circle inscribed in the rhombus $MEPF$. We have that
$$Area(MEPF) = \frac{1}{4} Area(ABCD) = 5,$$
and
$$ME = \frac{1}{2}MC = \frac{1}{2}\sqrt{MB^2 + BC^2} = \frac{5}{2}.$$
Thus, it follows that
$$r = \frac{Area(MEPF)}{\frac{1}{2}(ME + EP + PF + MF)} = 1.$$

□

Problem 4.296. *Given 25 chairs around a round table, such that on each chair exactly one person is sitting. After the break each of them can sit on the fifth chair that comes after the chair he/she was sitting before the break (toward either direction). The counting is done starting from the next chair. On each chair is again sitting exactly one person. After the break, in how many different ways can these 25 people sit on these 25 chairs?*

(A) 1024   (B) 32   (C) $(5!)^5$ (D) 5!   (E) 120

Solution. Answer. (B)
Let us enumerate these chairs from 1 to 25 toward a certain direction.
Note that people who were sitting on chairs 1, 6, 11, 16, 21 before the break, then after the break they sit again on these chairs.
$i \to j$ means that before the break the person who was sitting on the ith chair, after the break is sitting on the jth chair.
They can do that in the following two ways:
either $1 \to 6 \to 11 \to 16 \to 21 \to 1$,
or $1 \to 21 \to 16 \to 11 \to 6 \to 1$.
Note that 25 chairs can be divided into five such groups, therefore after the break these 25 people can sit on the chairs around the round table in $2^5 = 32$ different ways. $\square$

Problem 4.297. *Given a triangle ABC, such that $AB < BC$. Let D be the midpoint of arc AC of the circumcircle of triangle ABC, such that points B and D are on different sides of the line AC. Let segment DE be perpendicular to chord BC. Given that $BE = 17$ and $EC = 7$. What is the length of side AB?*

(A) 7   (B) 8   (C) 9   (D) 10   (E) 11

Solution. Answer. (D)
Given that $\angle ABD = \angle CBD$ (see the figure).

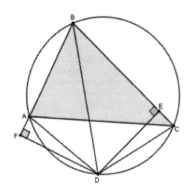

Let $DF \perp AB$, then $DE = DF$. Besides, $\angle DAF = \angle DCE$, therefore triangles DAF and DCE are congruent triangles.
Taking this into consideration, we deduce that
$$AF = CE = 7.$$

On the other hand, triangles BED and BFD are congruent triangles.
Thus, it follows that
$$BF = BE = 17.$$

Hence, we obtain that
$$AB = BF - AF = 17 - 7 = 10.$$

$\square$

Problem 4.298. *Given that the angle measures of each of $\dfrac{n^2}{100}$ angles of a convex n–gon is equal to n. What is the total number of all possible values of n?*

(A) 0   (B) 1   (C) 10   (D) 11   (E) infinitely many

Solution. Answer. (B)

According to the assumption of the problem, we have that $\dfrac{n \cdot n}{100}$ is a positive integer. Thus, it follows that $n = 10k$, where k is an integer.

We have that
$$\frac{n \cdot n}{100} \leq n.$$

Hence, we obtain that $k \leq 10$.

Note that the case $k = 10$ is impossible, as $100 \cdot 100° \neq 180° \cdot 98$.

Note also that the case $k = 9$ is impossible, as the sum of the exterior angles of a convex n–gon is equal to $360°$.

If $k \in \{1, 2, ..., 8\}$, then according to the fact that at most three angles of a convex of a convex n–gon can be acute. We obtain that $k = 1$. Therefore, the total number of all possible values of n is equal to 1. $\square$

Problem 4.299. *Positive integer n is called "strange" if numbers $d(n)$, $d(n^2)$, $d(n^3)$ form an airthmetic sequence, where $d(m)$ is the total number of all positive integer divisors of a positive integer m. What is the total number of all two-digit "strange" numbers?*

(A) 10   (B) 12   (C) 21   (D) 28   (E) 32

Solution. Answer. (D)

Note that if $n = p^{\alpha}$, where p is a prime number and α is a positive integer, then the numbers of divisors of n, n^2, n^3 are $\alpha + 1$, $2\alpha + 1$, $3\alpha + 1$, respectively. Moreover, these three numbers form an arithemtic sequence.

Now, let us prove that if a two-digit number n is not of the form $n = p^{\alpha}$, then it is not a "strange" number.

If $n = p_1^{\alpha_1} p_2^{\alpha_2}$, where $p_1 < p_2$ and p_1, p_2 are prime, α_1, α_2 are positive integers, then the numbers of divisors of n, n^2, n^3 are respectively equal to:
$$(\alpha_1 + 1)(\alpha_2 + 1), (2\alpha_1 + 1)(2\alpha_2 + 1), (3\alpha_1 + 1)(3\alpha_2 + 1).$$

Note also that
$$(\alpha_1 + 1)(\alpha_2 + 1) + (3\alpha_1 + 1)(3\alpha_2 + 1) > (2\alpha_1 + 1)(2\alpha_2 + 1).$$

Thus, it follows that n is not a "strange" number.

If $n = p_1^{\alpha_1} p_2^{\alpha_2} p_3^{\alpha_3}$, then in a similar way one can prove that n is not a "strange" number.

Hence, all possible values of n are:
$$2^4, 2^5, 2^6, 3^3, 3^4, 5^2, 7^2, 11, 13, 17, 19, 23, 29, 31, 37, 41, 43, 47, 53, 59, 61, 67, 71, 73, 79, 83, 89, 97.$$

Therefore, the total number of all two-digit "strange" numbers is equal to 28. $\square$

Problem 4.300. *Let n be a positive integer greater than 3. Given that numbers 1, 2,..., n can be divided into several groups, such that the greatest element of each group is equal to the sum of all the other elements of that group. What is the smallest possible value of n?*

(A) 10   (B) 11   (C) 12   (D) 13   (E) 14

Solution. Answer. (C)
Let us consider the following cases.
Case 1. If the number of the groups is equal to 1.
In this case, we obtain that
$$1 + 2 + \ldots + n = 2n,$$
where $n = 3$, which is impossible.
Case 2. If the number of the groups is equal to 2, then $n \geq 6$ and
$$1 + 2 + \ldots + n = 2n + 2a,$$
where n and a are the greatest numbers of these groups. Therefore
$$\frac{n(n+1)}{2} \leq 2n + 2(n-1).$$
Hence, we obtain that
$$n(n-7) + 4 \leq 0.$$
Thus, it follows that $n = 6$ and $9 = 2a$, which is impossible.
Case 3. If the number of the groups is equal to 3.
In this case $n \geq 9$ and
$$1 + 2 + \ldots + n = 2n + 2a + 2b,$$
where n, a and b are the greatest numbers of these groups.
Hence, we obtain that
$$\frac{n(n+1)}{2} \leq 2n + 2(n-1) + 2(n-2).$$
Thus, it follows that
$$n(n-11) + 12 \leq 0.$$
Therefore $n = 9$, which is impossible as in this case we obtain that $27 = 2a + 2b$.
Case 4. If the number of the groups is not less than 4.
In this case, we have that $n \geq 12$.
Note that the smallest possible value of n is 12, for example:
$$(12, 11, 1), (10, 4, 6), (8, 3, 5), (9, 7, 2).$$

□

Bibliography

[1] MAA American Mathematics Competitions *AMC 10*

[2] Sedrakyan H., Sedrakyan N., *AMC 12 preparation book*, USA (2021)

[3] Sedrakyan H., Sedrakyan N., *AMC 8 preparation book*, USA (2021)

[4] Sedrakyan H., Sedrakyan N., *Number theory through exercises*, USA (2019)

[5] Sedrakyan H., Sedrakyan N., *How to prepare for math Olympiads*, USA (2019)

[6] Sedrakyan H., Sedrakyan N., *The Stair-Step Approach in Mathematics*, Springer Int. Publ., USA (2018)

[7] Sedrakyan H., Sedrakyan N., *Algebraic Inequalities*, Springer Int. Publ., USA (2018)

[8] Sedrakyan H., Sedrakyan N., *Geometric Inequalities. Methods of proving*, Springer Int. Publ., USA (2017)

Made in United States
Troutdale, OR
09/22/2024